Turbulence and Diffusion in the Atmosphere

Springer-Verlag Berlin Heidelberg GmbH

Alfred K. Blackadar

Turbulence and Diffusion in the Atmosphere

Lectures
in Environmental Sciences

With 49 Figures
and Two MS-DOS Program Diskettes

 Springer

Professor Alfred K. Blackadar

The Pennsylvania State University
Department of Meteorology
Room 503 Walker Building
University Park, PA 16802
USA
e-mail: akb1@psu.edu

Cover picture: An artificially produced preconvective cumulus cloud caps the rising column of heated gases in this photograph of the mine-mouth Keystone Generating Station situated in western Pennsylvania. It was taken by Mr. Francis Schiermeier of the United States Environmental Protection Agency on 1 May 1969, a day with very light wind. In order to make the effluent gases visible for this photograph, the electrostatic precipitators of this 1800 MW power plant were temporarily suppressed.

Additional material to this book can be downloaded from http://extras.springer.com

Library of Congress Cataloging-in-Publication Data.
Blackadar, Alfred K., 1920 – . Turbulence and diffusion in the atmosphere: Lectures in Environmental Sciences / Alfred K. Blackadar. p. cm. Includes bibliographical references and index.
1. Atmospheric turbulence. 2. Atmospheric diffusion. I. Title.
QC880.4.T8B54   1997   551.55–dc20   96-35289

ISBN 978-3-642-64425-2   ISBN 978-3-642-60481-2 (eBook)
DOI 10.1007/978-3-642-60481-2

Typesetting: Data conversion by Steingraeber, Heidelberg
Cover design: Erich Kirchner, Heidelberg
Production Editor: P. Treiber

SPIN 10517839   55/3144 – 5 4 3 2 1 0 – Printed on acid-free paper

Preface

This book grew out of an introductory course that I was invited to teach on a number of occasions to senior and graduate level students at the University of Kiel. I have cherished these opportunities in part because I was never required to conduct examinations or give grades. For the students, however, my good fortune presented special problems that induced my sympathy: in addition to having to contend with a foreign language, they would eventually have to confront an examiner with his own ideas about what they should have learned. Although I always left a copy of my lecture notes with this person, they were too sketchy to be of much use. The present book is an attempt to solve some of these problems.

The content is intended to be as broad as possible within the limitations of an introductory one-semester course. It aims at providing an insightful view of present understanding, emphasizing the methods and the history of their development. In particular I have tried to expose the power of intuitive reasoning – the nature of tensor invariants, the usefulness of dimensional analysis, and the relevance of scales of physical quantities in the inference of relationships. I know of no other subject that has benefited more from these important tools, which seem to be widely neglected in the teaching of more fundamental disciplines.

This text is intended to be used by students who have had no previous experience in the subject. An introductory knowledge of hydrodynamics is assumed. The text is directed mainly for the use of students of meteorology and other specializations of atmospheric science. The problems and examples have been selected with the intention of broadening the student's understanding of the atmosphere, as well as his or her understanding of basic mathematics and the scientific method.

Many disciplines have left their mark on the subject, and the subject in turn has left its mark in the solution of important applications related to the structure of the atmospheric boundary layer and methods of predicting the impact of industrial activities. These problems are so broadly interdisciplinary that even the so-called experts have difficulty integrating them without the use of computational models. In the final appendices I have briefly discussed two such models which students can run experimentally on a PC, to see at first hand the usefulness of the concepts they will hopefully have gained from their studies.

Many friends and colleagues have provided stimulation and encouragement that made this book possible. I wish particularly to thank Professors Lutz Hasse, Rolf Kaese, John Wyngaard, and Dan Seidov for their valuable suggestions and critical reviews of the manuscript. I am indebted to Dr. Dennis A. Trout, of the U.S. Environmental Protection Agency for his careful reading of Chap. 10 and his helpful suggestions. Many typographical mistakes and errors in equations were corrected as a result of careful reading by Clemens Simmer, Hans-Jörg Isemer, and Sergei Gulev. I thank Ms. Frauke Nevoigt for her skillful redrawing of the majority of the illustrations and Mr. Francis Schiermeier of the U.S. Environmental Protection Agency for permission to use the cover photograph.

October 1996   *A. K. Blackadar*

Contents

1 The Nature of Turbulence

Turbulence comprises a large class of motions with complex, irregular and rather unpredictable natures. An unambiguous definition of the phenomenon is easily arrived at in some reasonably simple flow configurations such as for example, flow in a pipe. However, given the enormous number and variety of degrees of freedom that exist in real-world flows, it has not been possible to arrive at a consistent definition of the phenomenon that can be universally applied to separate turbulence from other complex types of flows, such as waves and large-scale circulation normally associated with weather patterns.

As an example of the difficulty, we are accustomed to thinking of the motions we observe meteorologically by averaging over a one- or five-minute period as an ordered field of motion that can be depicted on weather maps or predicted with computer models. By contrast, the myriad of gust eddies that influence the wind vane from second to second are regarded as turbulence, the effects of which can be dealt with only in a statistical way. However, climatologists prefer to study motions that are averaged over a period of many years, with the consequence that the day-to-day changes become the superimposed turbulence in their way of thinking. An important question is whether it is possible to set an absolute criterion to isolate turbulence as a phenomenon that follows a uniform set of laws.

Although no definition of turbulence can be given at this time, there is widespread agreement concerning some of its attributes. Lumley and Panofsky (1964) have drawn attention to the following points.

1. Turbulence is stochastic by nature. Even though turbulent motions are subject to deterministic equations, we recognize that these equations are nonlinear by nature. As a result the future characteristics of the motions are highly sensitive to small differences in the initial state, and it is not possible to observe the initial state sufficiently accurately for us to be able to treat the turbulent motions in a deterministic way.
2. Turbulence is three-dimensional. Although it is possible to discuss two-dimensional eddies, such as, for example, cyclones and anticyclones in the atmosphere's general circulation, their ensemble behavior is not similar to that of small-scale turbulence in a large three-dimensional environment.
3. Any two marked particles that are free to move within a turbulent environment tend to become increasingly distant from each other as time goes on.
4. Turbulence is by nature rotational. Vorticity is an essential attribute.

5. Turbulence is dissipative. The energy of turbulence tends to shift from large, well-organized eddies (small wavenumbers) toward smaller eddies and eventually into molecular motions. Because of 2, 3, and 4, vortices tend to be stretched by turbulence with a corresponding reduction in their diameter.
6. Turbulence is a phenomenon of large Reynolds numbers (i.e., large spatial dimensions and small viscosity). The space available for the motions must be large compared to the dimensions of eddies that are quickly dissipated.

These attributes can be observed by shedding a drop of ink into a pan of water that has been allowed to settle quietly for a few minutes. Due to the insertion momentum and density differences, turbulence is generated in the vicinity of the drop, which is readily seen, especially if good lighting is available. At first one sees that the initially concentrated blob begins to separate into a number of blobs connected by fine dark filaments, which become more and more elongated within a few seconds. As the volume spreads further, the dense filaments disappear, and the whole mass gradually enlarges itself becoming more and more diffuse until eventually it becomes unidentifiable. It is quite readily seen that all of the attributes enumerated above are present in this evolution. The filaments that form initially constitute vortex lines that elongate as time goes on and in doing so concentrate their vorticity (and marked fluid particles) into ever thinner diameters, until somewhat later molecular motions cause the spinning structure to be dissipated and to disappear.

Additional insight into the phenomenon can be gained by repeating this procedure using a flat dinner plate covered with a thin layer of water to which a small amount of detergent has been added to enable it to wet the surface uniformly. The ink drop in this case simply spreads out rapidly into a uniform smear a few centimeters in diameter and comes to a halt. The thinness of the fluid requires all but microscopic vortices to be two-dimensional, and viscosity is so powerful that the microscopic vortices, if they form at all, are almost instantly dissipated. The process that forms the filaments does not occur, and even the two-dimensional horizontal motions very quickly succumb to the effects of viscosity.

1.1 Two-Dimensional Eddies in the Atmosphere

Most, if not all, atmospheric motions are turbulent to some degree. Using terms that we shall define rigorously later on, we can talk of the turbulent *intensity* as the ratio of the energy associated with turbulence to that associated with the nonturbulent quasi-steady or mean motion. Since no unique definition of turbulence can be given, it follows that there is no unique way of separating the total motion. For an airplane pilot, the mean motion is that of the wind he or she uses to navigate, while turbulence is what shakes the airplane. For a climatologist, however, the mean wind is an average of 30 years of daily observations, and the superimposed variations are represented by arrows on a wind rose.

In 1921, A. Defant suggested that the migratory cyclones and anticyclones, and the role they play in the global distributions of wind, temperature and precipitation might be fruitfully studied by treating them as large-scale turbulence superimposed on the mean general circulation. In this way it was hoped that principles of turbulent transfer, as well as the limited understanding of the origin and behavior of turbulence that was actively emerging about this time might be applied to the better understanding of climate and daily weather changes. If one accepts this picture of the large-scale migratory flow patterns, then the high- and low-pressure systems depicted in Fig. 1.1 and their associated wind circulations constitute turbulence, while the mean circulation is characterized by the well-known quasi-permanent low-pressure centers located around 60° and the corresponding high centers around 30° latitude.

The behavior of these vortices superimposed on the semipermanent or mean circulation display many attributes commonly associated with turbulence. They are responsible for a large part of the heat and water vapor that are transported each year from the equatorial to higher latitudes. They also play a significant role in transporting the excess angular momentum generated in the surface layers of the atmosphere in the tropics to middle latitudes where it is dissipated by the frictional drag of the westerlies. Until the 1940s it was believed that the energy that drives the mean westerly circulation of middle latitudes was generated in low and high latitudes by the direct meridional cells in the general circulation (Rossby, 1941). The waves and circulations that developed in middle latitudes were viewed as the products of instability of the westerlies, which derived their energy from the mean westerlies and thus ultimately from the direct meridional circulations in low and high latitudes.

This simple theory of middle-latitude vortices changed drastically at the end of that decade as a result of the work of Starr (1948, 1951) and Blackadar (1950). It is now known that available potential energy is converted into kinetic energy primarily in the scale of the cyclonic and anticyclonic vortices, from which it is transformed into the kinetic energy of the larger, global-scale circulation. This behavior was at first difficult to grasp as it appeared to violate the principles of the second law of thermodynamics, which has been widely interpreted as requiring large-scale organized circulations to decay into smaller ones. Certainly it is inconsistent with the behavior described by number 5 in the list above.

The conversion of relatively small two-dimensional eddies into larger eddies and/or into the mean circulation can readily be observed in a coffee cup or shallow pan of water. By stirring with a spoon, one can create a small vigorous two-dimensional eddy near the edge of the container. When left to itself, such an eddy moves toward the center of the container, and, in so doing, changes into a larger mean-circulation eddy that fills the entire container. On weather maps such as shown in Fig. 1.1, developing cyclones, forming along a front in middle latitudes, tend to move northeastward as they grow and intensify; likewise, migratory high centers tend to move southward and merge with the semipermanent high-pressure cells near latitude 30°. The result of this characteristic behavior is that the energy

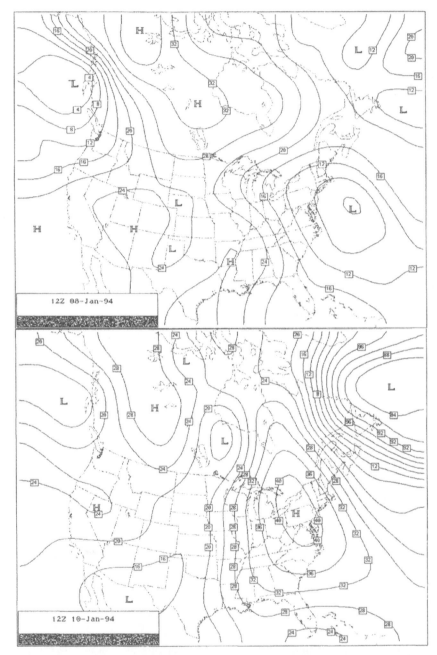

Fig. 1.1 Two surface-pressure maps for 8 and 10 January 1994 over North America and neighboring oceans. The high-pressure system in western Canada moves southeastward and intensifies, while the coastal low-pressure center moves northeastward and also intensifies. The wind systems surrounding these centers represent intensifying vortices. The motion of the centers over the two-day interval is such that the kinetic energy, which initially was associated with relatively small migratory eddies, is becoming transformed into that of the global mean circulation

of the smaller migratory cells is converted into a pattern that is identified with the global larger-scale circulation pattern.

These migratory eddies are basically two-dimensional in nature. The mean vertical motions in these systems are of the order of 1 part in 1000 of the horizontal wind speeds. Thus these types of eddies also fail to satisfy the second attribute of the above listing. Attributes 2 and 5 are closely related. The cascade of energy from larger to smaller vortices is believed to be accomplished by vortex stretching, a process that is impossible with two-dimensional motions.

There is no universally accepted definition of turbulence, and there are numerous references to what is termed *two-dimensional turbulence*, particularly in the oceanographic community. It should be emphasized, however, that two- and three-dimensional turbulence phenomena are fundamentally different in nature. Unless specifically indicated to the contrary, these lectures will be concerned with three-dimensional turbulence.

1.2 The Reynolds Number and Its Significance

Since the Reynolds number is fundamental to the existence of turbulence, we shall begin by looking at its definition and meaning.

We can imagine that a motion-picture director, faced with the need to film the conflagration of a large structure, might attempt to save costs by building a small match-box sized replica and then filming the conflagration with a close-up lens, perhaps in slow-motion. Anyone who has burned a matchbox knows that the result would not be satisfactory. The full-scale conflagration is marked by an enormous range of flame sizes, while in the model only a narrow spectrum of sizes, or perhaps only a single flame, can be found. The very small eddies and temperature differences needed to support the large range of sizes in the full-scale structure are dissipated too quickly by viscosity and heat conduction because of the sensitivity of these processes to physical length scales. The Reynolds number can be defined as the ratio of two lengths: a length L characteristic of the dimensions of the space available for all scales of motion, and another length that measures the thickness of the so-called laminar or viscous sublayer – a layer so thin that turbulence cannot be maintained, even when it is present in other parts of the fluid.

The length L is normally defined by the boundary configuration – the depth of the water in a channel, or the diameter of a confining pipe. In the atmosphere, gravity plays a strong role. As a result of its influence, L is affected by the temperature gradient and turbulence becomes restricted by stable lapse rates as well as by physical boundary constraints. We shall consider this situation later.

The laminar sublayer thickness δ is related to viscosity. Clearly, it is the kinematic, rather than the dynamic viscosity that concerns us, since its dimensions are independent of mass. To form a length out of the kinematic viscosity it is necessary to divide by a velocity. For this we can choose the velocity U of the largest scale of motions. (The velocity of any other scale of eddies in the spectrum can be expected to vary in proportion to that of the mean free-stream velocity.)

$$\delta \propto \nu/U \tag{1.1}$$

where ν denotes the kinematic viscosity. Some typical values of ν are listed in Table 1.1.

The *Reynolds number* is the dimensionless ratio of these two lengths. Its magnitude may be thought of as the ratio of the size of the largest energy-containing eddies to the size of the eddies mainly responsible for the dissipation of kinetic energy into heat.

$$\text{Re} = \frac{L}{\delta} = \frac{LU}{\nu}. \tag{1.2}$$

One occasionally observes transitions from laminar to turbulent flow. Examples are the slow opening of a water faucet and the smoke ascending from a lighted cigarette in a stuffy room. There are always ambiguities attending the definition of L and U in such situations. In pipes one can naturally look at the diameter and mean current velocity. In this case the transition to turbulent flow occurs with a Reynolds number of the order of 1000. Laminar flows may exist at higher values, but become sensitive to the presence of disturbances. Turbulence can occur with lower values, but tends to die out. Other transition phenomena, like the smoke ascending from a cigarette, seem to occur with critical values closer to unity.

Table 1.1 Kinematic viscosity of various natural substances at sea level

Substance	Temperature °C	$\nu \; m^2 s^{-1} \times 10^{-6}$
Air	0	13
	20	14.8
	40	16.8
Water	0	1.79
	20	1.01
	40	0.66
Ice (glacial)	...	1.4×10^{16}

1.3 The Reynolds Approach
to the Equations of a Turbulent Fluid

The quantitative approach to the study of turbulent fluids was begun in 1895 by Osborne Reynolds. This classic paper is widely available in libraries. It is beautifully written, easy to understand, and should be required reading for every student. Reynolds' method has dominated the quantitative treatment of turbulent fluids up to the present time. His study was restricted to incompressible fluids;

later this restriction was removed by L. F. Richardson (1920). Richardson is also well known as a pioneer in the development of numerical weather prediction.

Reynolds separated each of the velocity components, which we denote by u, v, and w, into two parts: a mean value, denoted by $\overline{u}$, $\overline{v}$ and $\overline{w}$; and a turbulent portion, denoted by u', v', and w'. From the definition, one has

$$u = \overline{u} + u' , \quad v = \overline{v} + v' , \quad w = \overline{w} + w' \tag{1.3}$$

The averaging process can be defined in many different ways, the method used being generally picked to fit the application at hand. For example when statistically steady state (stationary) flows in a pipe or wind tunnel are being examined, it is natural to observe a sequence of velocities of different particles at times that are sufficiently separated that they may be considered to be uncorrelated with each other. The mean value of such a sequence converges to a value that is independent of the values of other samples. Moreover, in such samples, the mean values of the deviations from the mean, i.e., u', v', w', are zero. In this case, the following relations are satisfied:

$$\overline{\overline{u}} = \overline{u}, \quad \overline{u'} = 0, \quad \overline{\overline{u}\,\overline{v}} = \overline{u}\,\overline{v} , \quad \overline{\overline{u}u'} = 0 ,\dots . \tag{1.4}$$

Averages of this kind are called ensemble averages; the average is taken over a large number of different independent samples of a flow determined by what are perceived to be the same initial and boundary conditions. A closely related type of average can be calculated from the continuous record of the flow at the same location over a period of time. The flows are said to be *ergodic* if these two means are the same, and it has been shown that a necessary condition for ergodicity is that the mean flow be independent of time (i.e., *stationary*) and location (i.e., *homogeneous*). Students should note these terms as they are often used in describing the turbulent environment.

It is also possible to define an average at a fixed time over a specified area or volume, or even to average over both a specified volume and over a specified time. Since turbulence, from the Reynolds point of view, is conceived as being the departure from the mean flow, the ambiguity of a definition of the mean flow is a serious impediment to a unique definition of the phenomenon of turbulence.

A completely homogeneous and stationary environment is not very interesting, nor is it commonly found in the natural atmosphere. Reynolds recognized this fact and proposed to deal with such flows by taking a running average over a specified time or space interval. In this way the mean and eddy values of the velocity components may be considered to be continuous functions of space and time. However, in this case, Reynolds recognized that the relations (1.4) cannot be exact. He reasoned that, unlike u, v, w and u', v', w', the values of $\overline{u}$, $\overline{v}$, $\overline{w}$ change rather slowly. Thus, the relations (1.4), which apply exactly in homogeneous, stationary conditions, could still be an acceptable approximation for the study of turbulent fluids. This assumption and the relations that result from it are known as the Reynolds postulates.

The Reynolds method consists of averaging each of the equations of motion and the equation of continuity. Reynolds dealt only with incompressible fluids, but this is not a serious restriction for many important problems. Richardson later applied the same methods to compressible flows, thereby opening up this method to the application of thermodynamic equations to turbulent phenomena. The general procedure is to replace u, v, w in each of the equations by $\bar{u}$, $\bar{v}$, $\bar{w}$, and u' v', w' using (1.1). The resulting equations are then averaged and simplified using the Reynolds postulates. The result is a set of equations for the changes of $\bar{u}$, $\bar{v}$, and $\bar{w}$.

1.4 Averaging the Equation of Continuity

We write the equation of continuity in the following forms, where the operator d/dt represents the rate of change following the instantaneous motion, and the partial derivatives represent the spatial and time derivatives at a fixed point. We refer to the former as the *Lagrangian* derivative and latter as the *Eulerian* derivative, corresponding to the two fundamental coordinate systems used to study fluid dynamics.

$$\frac{d\rho}{dt} + \rho\frac{\partial u}{\partial x} + \rho\frac{\partial v}{\partial y} + \rho\frac{\partial w}{\partial z} = 0 \,, \tag{1.5}$$

$$\frac{\partial\rho}{\partial t} + u\frac{\partial\rho}{\partial x} + v\frac{\partial\rho}{\partial y} + w\frac{\partial\rho}{\partial z} + \rho\frac{\partial u}{\partial x} + \rho\frac{\partial v}{\partial y} + \rho\frac{\partial w}{\partial z} = 0 \,. \tag{1.6}$$

We shall frequently use the index notation in writing equations like this, letting a subscript i represent any of the three Cartesian coordinate directions. We include the convenient Einstein summation convention, which prescribes that if any of the indices is repeated in a term, a summation over that index is implied[1]. Equation (1.6) may thus be written as

$$\frac{\partial\rho}{\partial t} + u_k\frac{\partial\rho}{\partial x_k} + \rho\frac{\partial u_k}{\partial x_k} = 0 \,. \tag{1.7}$$

We can note that since the implied summation is over all three values of the index k, k is a dummy index which requires no special definition. Since the latter two sets of terms constitute the expansion of the derivative of a product, we can write (1.7) simply as

$$\frac{\partial\rho}{\partial t} + \frac{\partial\rho u_k}{\partial x_k} = 0 \,. \tag{1.8}$$

We now employ the Reynolds procedure by substituting from (1.3) and then averaging the result:

---

[1] The reader should note that square of an indexed quantity represents an implied repetition of the index and is therefore always to be summed. For example $u_i^2 = u_i u_i$.

$$\frac{\overline{\partial \rho}}{\partial t} + \frac{\overline{\partial \rho u_k}}{\partial x_k} = 0 . \tag{1.9}$$

The bar is purposely extended over the derivatives to indicate that it is the derivatives that are averaged. However, since the averaging process is a linear summation process, the derivative of a sum is equal to the sum of the derivatives. We also invoke the Reynolds postulates so that ρ is replaced by $\overline{\rho}$ and $\overline{\rho u_k}$ is replaced by $\overline{\rho}\ \overline{u}_k$. The final result is the equation

$$\frac{\partial \overline{\rho}}{\partial t} + \frac{\partial \overline{\rho}\ \overline{u}_k}{\partial x_k} = 0 . \tag{1.10}$$

Our analysis has shown that the equation of continuity can be applied without change to the mean motion and to the mean density. We must, however observe some caution when we retrace our steps back to the form of (1.5) and (1.6). In place of (1.6), we have

$$\frac{\partial \overline{\rho}}{\partial t} + \overline{u}_k \frac{\partial \overline{\rho}}{\partial x_k} = -\overline{\rho}\frac{\partial \overline{u}_k}{\partial x_k} . \tag{1.11}$$

The terms on the left-hand side are no longer a simple substantial derivative, but rather the change that one observes in a coordinate system that moves with the mean velocity. This system is no longer Lagrangian in the sense that we follow a particular particle. Following individual, tagged particles is an exceedingly difficult task, and the small number of such observations is one of the greatest problems we have in trying to construct models for predicting diffusion.

1.5 Fluxes and the General Conservation Equation

Consider a certain class of fluid properties that are often dealt with in physics, viz. properties whose amounts are identified with a mass of the fluid. Such a property might be specific humidity (the mass of water vapor per unit mass of air), the kinetic energy per unit mass, or the specific entropy ($c_p \ln \theta$). We would not include in this class the potential temperature θ itself, because this is not in any way associated with the amount of fluid that is present. We shall also confine our consideration to conservative properties – properties that do not change just because of their motion. Therefore, we would not include enthalpy ($c_p T$) because the temperature changes when the pressure changes. Properties of the acceptable kind that we shall frequently deal with are the water vapor mass concentration, momentum, kinetic energy, and pollutant concentrations.

We let q stand for any such property per unit mass. The equation that expresses the conservation of such a property is

$$\frac{\mathrm{d}q}{\mathrm{d}t} = S_q . \tag{1.12}$$

In this equation the symbol S_q refers to the source strength of the quantity per unit mass. In the case of water vapor, it might refer to the rate of evaporating of liquid water, for example. We can expand this equation to

$$\rho\frac{\partial q}{\partial t} + \rho u_k \frac{\partial q}{\partial x_k} = \rho S_q \ . \tag{1.13}$$

We now multiply the equation of continuity term-by-term by q,

$$q\frac{\partial \rho}{\partial t} + q\frac{\partial \rho u_k}{\partial x_k} = 0 \ , \tag{1.14}$$

and add the two equations together:

$$\frac{\partial(\rho q)}{\partial t} = \rho S_q \ - \ \frac{\partial(\rho u_k q)}{\partial x_k} \ . \tag{1.15}$$

On the left side of this equation we have the rate of change of the amount of q per unit volume measured at a fixed location. The first term on the right gives the rate of internal production of q per unit volume at the same fixed point. The last term represents the rate of convergence of a vector whose components are $(\rho u_k q)$. What is the physical meaning of this vector?

We let A stand for a small area on the surface perpendicular to any one of the three axes. It is evident that the product u_kA is the volume of a cylinder with a height of $u_k \times$ (unit time), with A its base, and accordingly ρu_k represents the mass transported through a unit area of the surface perpendicular to x_k, in one unit of time. Multiplication of ρu_k by q, the amount of the property per unit mass of fluid then represents the amount of the property transported per unit area and per unit time across a surface perpendicular to x_k. We call this quantity the flux of q in the x_k direction. The three terms considered together as the components of a vector represent the flux of q through a surface normal to the velocity vector.

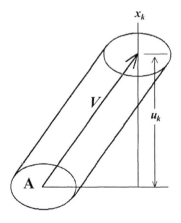

Fig. 1.2 The flux vector can be calculated by noting that all of the fluid contained in the oblique cylinder is transported through the surface A in a unit time

Looking again at (1.15), we interpret the rightmost term as the convergence of the flux vector of q. By Gauss's theorem, this convergence is the amount of the quantity q that enters the unit volume per unit time from sources outside its boundaries.

We would like to replace the variables in (1.15) with mean quantities. To this end, we average the terms as before and employ the Reynolds postulates. In doing so, we encounter the new product $\overline{\rho q u_k}$. Since $\overline{\rho' q u_k}$ is zero, the new product is equivalent to $\overline{\rho}\,\overline{q u_k}$. Expanding $\overline{q u_k}$ we get

$$\overline{(\overline{q} + q')(\overline{u}_k + u'_k)} = \overline{q}\,\overline{u}_k + \overline{q'\overline{u}_k} + \overline{\overline{q}u'_k} + \overline{q'u'_k} \ . \tag{1.16}$$

The middle two terms on the right are zero according to the Reynolds postulates. However, the last term on the right, the covariance of the two fluctuating quantities, is not generally zero. Thus we see that the mean value of the flux consists of two parts: a flux determined by the mean values of the fluctuating quantities, and the so-called eddy flux produced entirely by the fluctuating quantities.

The averaged conservation equation then becomes

$$\frac{\partial \overline{\rho}\,\overline{q}}{\partial t} = \overline{\rho S_q} - \frac{\partial(\overline{\rho}\,\overline{u}_k \overline{q})}{\partial x_k} - \frac{\partial\left(\overline{\rho}\,\overline{u'_k q'}\right)}{\partial x_k} \ . \tag{1.17}$$

We see that the external source of the mean property $\overline{q}$ consists of two parts, in general: a portion that is identified with the mean state quantities, and that due to the convergence of the eddy flux. We can obtain another common form of this equation by subtracting term-by-term the mean equation of continuity multiplied by $\overline{q}$, with the result

$$\overline{\rho}\frac{\partial \overline{q}}{\partial t} + \overline{\rho}\,\overline{u}_k \frac{\partial \overline{q}}{\partial x_k} = \overline{\rho S_q} - \frac{\partial}{\partial x_k}\left(\overline{\rho u'_k q'}\right) \ , \tag{1.18}$$

which can be rewritten in the form

$$\frac{\mathrm{D}\overline{q}}{\mathrm{D}t} \equiv \frac{\partial \overline{q}}{\partial t} + \overline{u}_k \frac{\partial \overline{q}}{\partial x_k} = \overline{S}_q - \frac{1}{\overline{\rho}}\frac{\partial \overline{\rho}\left(\overline{u'_k q'}\right)}{\partial x_k} \ . \tag{1.19}$$

The operator $\mathrm{D}/\mathrm{D}t$ that is used in this equation is the rate of change that occurs in a coordinate system that moves with the mean motion. Unlike the unaveraged substantial derivative, this system does not consist of a conserved mass. A system that consists of the same mass at all times is called a closed system. Most thermodynamic and physical laws are formulated for closed systems. Turbulent systems are open systems, and it becomes necessary to reconsider the familiar laws in order to make them applicable to the mean turbulent state.

1.6 The Closure Problem

Of the seven equations used in dynamical prediction models, six are applications of the conservation equation (three equations of motion, the equation of continuity, and the equations for conservation of water vapor and energy). All of these are nonlinear, and when averaged, they contain turbulent fluxes of the mean quantities. The result is twelve new variables in addition to the original seven – twelve more variables than there are equations to solve for them. Before the system can be closed, we must find new equations that relate the covariances to the mean quantities for which we already have equations. The final number of equations must equal the total number of unknowns.

The solution most often employed in the past is to try to relate the fluxes to the spatial derivatives of the mean quantities. In special cases, this procedure makes many of the new variables zero. This type of closure is known as first-order closure because only equations for the first moment (i.e. mean) quantities remain after the closure is complete.

It is quite straightforward to derive equations for the rate of change of the second-order statistics (i.e. the fluxes). When this is done, a distressing situation emerges. The equations for the second order quantities contain a new set of variables, the third moments, or mean values of triple products of turbulent quantities. One can continue the process by deriving equations for the third moments, but it is found that these contain fourth order moments, and so on, and on. Therefore, we must recognize the fact that there is no straightforward way to close the set of atmospheric equations. Instead, it is necessary to make hypotheses about these relations. When these hypotheses are introduced into the second order equations for the covariances, the scheme is referred to as second-order closure. The availability of high-speed computers has resulted in many attempts at *second-order closure* in recent years.

1.7 First-Order Closure – Exchange Theory

The basic ideas of first-order closure, or *K-theory* as it is often called, have been generally credited to Schmidt (1925) and Prandtl (1925), but according to Prandtl the principal concepts go back to Boussinesq (1897). These Austausch or exchange theories are quite similar to the kinetic theory of gases. The primed quantities of Reynolds are identified with individual fluid parcels which are the kinetic analogues of molecules. Just as the molecules exchange properties from time to time by collisions with each other, so the fluid parcels were envisioned as mixing with each other at more or less discrete intervals, between which events they characteristically move a certain distance, the so-called mixing length, analogous to a molecular mean free path. Despite its obvious shortcomings, this model provides a conceptual basis for relating the fluxes to the gradient of the mean quantities. It also enables us to identify some limitations of the theory.

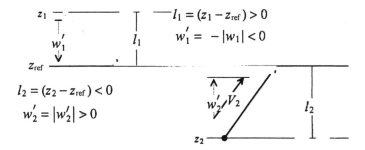

Fig. 1.3 Properties of the trajectories of two particles discussed in exchange theory

Refer to Fig. 1.3. Let z_{ref} refer to a reference level at which the vertical eddy flux of some property q, say the water vapor mass concentration, is to be estimated. We shall assume that the distribution of $\bar{q}$ in the neighborhood of z_{ref} is smooth, so that the value of $\bar{q}$ can be estimated by using the Taylor series expansion and neglecting terms of higher order than the first.

$$\bar{q}(z) = \bar{q}(z_{ref}) + (z - z_{ref})\frac{\partial \bar{q}}{\partial z} . \tag{1.20}$$

Consider a typical parcel moving downward through the reference level with a vertical velocity $-|w_1'|$, and let l_1 represent the distance above the reference level when it last mixed with its environment and thereby took on the mean value at that level. The deviation of its value from the mean value at the reference level is

$$q_1' = \bar{q}(z_1) - \bar{q}(z_{ref}) = |l_1|\frac{\partial \bar{q}}{\partial z} , \tag{1.21}$$

and its corresponding contribution to the flux is

$$w_1' q_1' = - |w_1'|\,|l_1|\frac{\partial \bar{q}}{\partial z} . \tag{1.22}$$

In like manner, a typical particle moving up from a lower level has a q value deviation of

$$q_2' = \bar{q}(z_2) - \bar{q}(z_{ref}) = - |l_2|\frac{\partial \bar{q}}{\partial z} \tag{1.23}$$

and a flux contribution equal to

$$w_2' q_2' = - |w_2'|\,|l_2|\frac{\partial \bar{q}}{\partial z} . \tag{1.24}$$

The coefficient that multiplies the gradient of $\bar{q}$ is always negative, so that when the average is taken over all the parcels crossing the reference level one obtains a relation

$$\overline{\rho w' q'} = -A\frac{\partial \bar{q}}{\partial z} = -\bar{\rho} K\frac{\partial \bar{q}}{\partial z} \tag{1.25}$$

in which A, the Austausch or exchange coefficient, and its kinematic analogue K, are positive definite quantities. A has dimensions $ML^{-1}T^{-1}$ (Mass × Length^{-1} × Time^{-1}), while K (which appeared in Boussinesq's form of the equation) has dimensions of L^2T^{-1}, or a length times a velocity. Since it contains no mass dimension, K is often called the *kinematic* exchange coefficient. It is also often called the *eddy diffusivity*. From above considerations and experience, we can identify L with the spatial size of the largest energy-containing eddies, and the velocity with the root-mean-square value of the turbulent velocities. The latter quantity is closely related to the turbulent energy, which we shall study later on.

As mentioned earlier, K-theory is modelled closely on the classical kinetic theory of gases. The length scale described above is usually called the mean free path, and is visualized as the mean distance traversed by a molecule between collisions. The K-theory analogue of the mean free path is usually called the *mixing length* or *Prandtl length*. Its nature is more difficult to define because the mixing process is not carried out by discrete events. About all one can do is use it as a measure of the size of the largest eddies, which are the ones responsible for most of the mixing that goes on.

A review of these ideas should remind us to confine our application of K-theory to quantities that are conservative for vertical motions. For example, under unsaturated conditions entropy depends only on potential temperature, which is conservative; but enthalpy, which depends on temperature is not conservative for vertical motions, and therefore will not be governed by K-theory. Water vapor is conserved in the absence of condensation, and experience has shown that the values of K that apply to the flux of water vapor and the flux of heat (as applied to the gradient of potential temperature) are probably about equal. K-theory has also been applied to momentum transfer, but there is considerable reason to doubt that momentum is completely conservative. Eddy motions are accompanied by pressure inequalities that influence them between mixing events. Experience shows that K values for momentum are not generally equal to K values for other properties, though they generally are of similar order of magnitude.

Unfortunately, since K depends on the statistics of turbulent motions, it is not a constant. Even if we could assume that the same K prevailed for all of the properties, we still would not have a complete closure of the equations. In later chapters we shall study methods by which the value of K can be estimated in certain cases.

1.8 Problems

1. The equation of continuity may be considered to be a special case of the general conservation equation for which q is 1. What is the physical meaning of q in this case? Carry out the derivation of the equation of continuity and discuss any assumptions that are required.

2. Consider the second term on the right side of the general conservation equation (1.17). Discuss the relative importance of advection by and divergence of the mean motion in changing the mean concentration of pollutants at a fixed point.

3. A typical value of the eddy diffusivity K in the surface layers is $1\,\text{m}^2/\text{s}$. This is about one hundred thousand times larger than the molecular diffusivity, in spite of the fact that typical molecular velocities are about 100 times larger than turbulent velocities in the atmosphere. In the light of the Prandtl/Schmidt closure, how can this be rationalized?

4. Name five flows in nature that are nonturbulent.

5. Would you expect the Great Nebula of Orion and other similar clouds of gas in space to be turbulent? What do pictures of these objects indicate?

6. Express the flux $\overline{\rho w'q'}$ in terms of the correlation coefficient r_{wq} between the fluctuations of w and q. Under what conditions would these quantities be expected to be correlated?

7. The entropy entering a system through its boundary is usually calculated by dividing the heat by the absolute temperature. Apply this relation in reverse to derive an expression for the mean turbulent heat flux in terms of the gradient of mean potential temperature, as follows:

(a) Write an expression for the turbulent flux of entropy.
(b) Make use of the approximation $(\ln \theta)' \cong \theta'/\overline{\theta}$, which is valid for $\theta' \ll \overline{\theta}$.
(c) Apply K-theory to relate the mean entropy flux to the gradient of mean potential temperature.
(d) Calculate the mean turbulent heat flux from the mean turbulent entropy flux, as described above. Why can we not follow the simpler procedure of relating the mean enthalpy flux to the gradient of mean temperature?

8. Which of the following expressions reduce to a scalar, and which to a vector?

(1) $p_{ik}\partial u_i/\partial x_k$

(2) $\varepsilon_{\iota\varphi\kappa}\Omega_\varphi v_\kappa$

(3) p_{ii}

(4) $l_{i\alpha}v_\alpha$

(5) $(K_{ijk} + K_{jik})\left(\dfrac{\partial u_j}{\partial x_k} + \dfrac{\partial u_k}{\partial x_j}\right)$

(6) $p_{ik}\delta_{ik}$

9. Although middle-latitude cyclones and anticyclones are not considered true turbulence, their activity in carrying heat northward from low to high latitudes has many of the characteristics we associate with turbulence. Using exchange theory, make an estimate of the order of magnitude of the average northward flux of heat in middle latitudes. The mean eddy flux of heat is defined as $c_p\overline{\rho v'T'}$, where c_p is the specific heat of constant pressure $(1004\,\text{J}\,\text{kg}^{-1}\text{K}^{-1})$, v' is the deviation from the average northward component of the wind, and T' the corresponding deviation of temperature from the mean.

2 The Navier–Stokes Equations

2.1 The Nature of Stress

A fluid parcel is influenced by its neighbors through forces that act across its bounding surfaces. Such forces tend to be proportional to the area of the surface across which they act. Therefore, it is logical to denote them by their value per unit area of that surface. This kind of force is called a stress.

Inherent in the designation of a stress is the orientation of the surface across which it acts. It can be shown that the stress acting across any surface with arbitrary orientation can be determined if the stresses acting across each of the three surfaces normal to the coordinate axes are known. There are nine such quantities: three components of the force for each of the surfaces, and three surfaces. We shall denote the total stress by p_{ik}. The first index i denotes the axis that is perpendicular to the surface, while the second index k indicates the component of the force vector acting across that surface.

Consider a surface in the xy-plane normal to z-axis, which we shall suppose to be directed upward. Let a tangential force be applied to the top of this surface in the x-direction. This stress is then given by the component p_{zx}, and its value is positive. The stress is also finite in magnitude, but it acts on a surface that has no volume and no mass! Therefore it must be that the fluid under the surface is exerting an equal and opposite force on the underside of the surface, so that the net force acting on the surface is zero. By convention the stress is defined by the force acting on the upper side of the surface, or more generally, that which acts on the positive side of the surface.

The ratio of surface area to volume increases without limit as the volume of a fluid parcel shrinks to zero, and this fact puts severe physical constraints on the field of stress. One of these is that the net force exerted on a closed surface must go to zero as the volume enclosed by the surface goes to zero. When this requirement is evaluated, there emerges a set of equations that govern the variation of each of the stress components as the axes of the coordinate system are rotated. We shall not go into details. However, we mention that arrays which satisfy these requirements are called tensors. Stress is a tensor of rank 2; vectors are tensors of rank 1; scalar quantities may be considered to be tensors of rank 0. The concept can be extended to higher ranks, and we shall encounter some of these later on.

Another property of stress that follows from the physical constraints is that the sum of the three diagonal components is a scalar quantity invariant with respect to any rotation of the axes. We might expect this sum to have a physical significance; its properties suggest that this is related to the pressure by

$$p \equiv -p_{ii}/3 \ . \tag{2.1}$$

The minus sign is required because the force exerted on the top (positive) side of the surface is in the negative direction. It should be mentioned that this is really only a plausible hypothesis; it is entirely possible that the pressure depends on other scalar quantities as well, like, perhaps the divergence of the velocity. That would seem to be inconsistent with the equation of state, but it must be remembered that the laws of thermodynamics have generally been developed in systems that are in equilibrium.

Not only must the net force shrink to zero as the volume goes to zero, the total moment of the forces must also approach zero. This requirement can be shown to result in the property called symmetry. As a result, columns and rows are interchangeable; the stress component p_{ik} is always equal to the component p_{ki}. As a result, there are only six independent components of the stress tensor.

When a fluid is at rest and in equilibrium, the stress consists only of the pressure, which is exerted equally in all directions. In a fluid that is in a state of motion, the stress field deviates from a simple pressure, but in terms of absolute magnitude, the pressure is still by far the largest component of the total stress. Since its behavior is governed by the laws of thermodynamics, it is customary to remove the pressure from the stress and treat it separately. The remainder is called the viscous stress. We shall see that the magnitude of a stress *per se* has no consequences for the motion, only its gradient. Small though it is, the viscous stress plays an important role in frictional and dissipative processes. We denote it by τ_{ik}, and its definition is

$$\tau_{ik} \equiv p_{ik} - \frac{1}{3} p_{jj} \delta_{ik} = p_{ik} + p \delta_{ik} \ . \tag{2.2}$$

In this equation, δ_{ik} is the unit symmetric tensor with a value of 1 if $i = k$ and 0 otherwise.

When the fluid is in equilibrium all the components of τ_{ik} are zero. The stress is then independent of the orientation of the coordinate system. The word that is used to describe this state is *isotropy*. An isotropic stress consists of three equal diagonal components, each of which is minus the pressure.

2.2 Invariants of Fluid Motions

Let us consider the motion of a fluid in the neighborhood of any selected point. One way to do this would be to expand the velocity components in a two or three dimensional Taylor series in powers of the components of the displacement from the point. If we stay sufficiently close to the point so as to keep these displacements small, we can neglect terms of higher order than the first. Thus

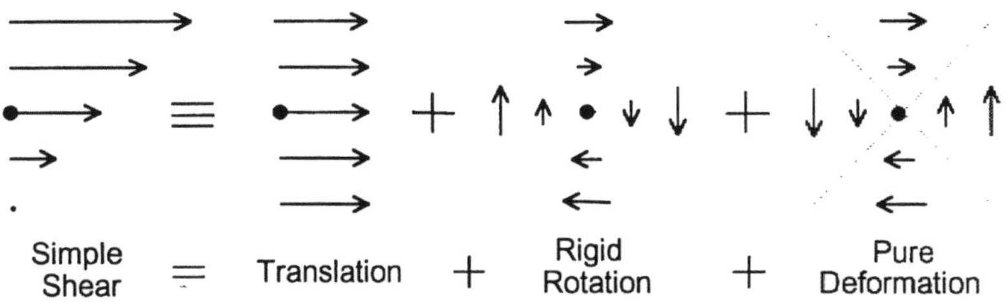

Fig. 2.1 Schematic decomposition of a simple shear flow into fundamental invariants

$$u_i(\delta x_k) = u_i(0) + \frac{\partial u_i}{\partial x_k} \delta x_k .\tag{2.3}$$

In order to understand better what we can learn from this exercise, we consider a simple two-dimensional flow consisting of a simple horizontal motion in the x-direction varying linearly with height (see Fig. 2.1). First we judiciously decompose the horizontal vectors that are functions of height only into three parts such that their vector sum at each level adds up to the original vector on the left side of the symbolic equal sign. The first of these consists of a uniform velocity equal everywhere to that of the reference point itself. This field is usually called the translation. Next we introduce two new fields of vertical vectors, depending linearly on x only, that are equal and opposite to each other, adding these fields to the second and third parts on the right side of the symbolic equation, as shown in the figure. We take care to see that the magnitude of the shear of these vertical vectors matches the magnitude of the horizontal shear in these two portions. The result of these operations is to create two two-dimensional motion types while guaranteeing that the sum of the three newly created fields is equal to the original field of motion. The first of the two-dimensional motion invariants is a rigid body rotation. The angular velocity of this motion is by definition one-half the vorticity. Any rigid body that is free to move with the fluid would partake of the first two parts of the motion. The last two-dimensional field making up the original field of motion is a non-rigid type called pure deformation. It consists of stretching along one direction and squeezing along a mutually perpendicular direction, in such a way that the divergence is zero. A fourth invariant, absent in this example, is the divergence; this is a scalar representing an expansion at a uniform rate in both the x- and z-directions.

These ideas can easily be extended to three dimensions. The translation and vorticity can be represented by three-dimensional vectors. The deformation rate can be described as stretching and/or squeezing along three mutually perpendicular directions in such a way that the three-dimensional divergence is zero. If we define a fluid parcel by a set of designated points making up its surface, then as the fluid moves, translation moves the volume as a whole, the vorticity causes it to rotate around a defined axis, divergence causes the volume enclosed within the boundary to expand at an equal rate in all directions, and pure deformation causes its shape to change.

We shall formalize this discussion by rewriting (2.3) as follows:

$$u_i\left(\delta x_k\right) \equiv u_i\left(0\right) + \frac{1}{2}\left(\frac{\partial u_i}{\partial x_k} - \frac{\partial u_k}{\partial x_i}\right)\delta x_k + \frac{1}{2}\left(\frac{\partial u_i}{\partial x_k} + \frac{\partial u_k}{\partial x_i}\right)\delta x_k . \qquad (2.4)$$

The first term is of course the translation. The second term contains the scalar product of the displacement vector and a tensor $1/2[(\partial u_i/\partial x_k)-(\partial u_k/\partial x_i)]$. This tensor is *antisymmetric* because interchanging the order of the indices reverses the sign of each of the components. Note that the three diagonal components of this tensor are zero. The scalar product represented by the second term can be shown to be minus one-half the ith component of the cross-product of the vorticity vector and the displacement vector, and it yields a velocity that would result from a rigid-body angular velocity vector equal to one-half of the vector vorticity. The third term is one-half the scalar product of the displacement vector with a tensor

$$D_{ik} \equiv \frac{\partial u_i}{\partial x_k} + \frac{\partial u_k}{\partial x_i} \qquad (2.5)$$

This tensor is symmetric and has a trace of twice the divergence:

$$D_{jj} = 2\frac{\partial u_j}{\partial x_j} = 2 \text{ div } V . \qquad (2.6)$$

The divergence represents the time rate of change of volume per unit volume of a given marked quantity of fluid and we do not want this included in the deformation rate. Therefore, to represent the rate of pure deformation we must subtract out the divergence. When we do so, we get

$$d_{ik} = D_{ik} - \frac{1}{3}\delta_{ik}D_{jj} = D_{ik} - \frac{2}{3}\frac{\partial u_j}{\partial x_j}\delta_{ik} . \qquad (2.7)$$

Any body subjected to this field of motion alone gets contracted or expanded in each of three mutually perpendicular directions subject to the condition that its volume does not change.

There is a powerful principle of physics that says that tensor invariants of one particular kind can only depend on other tensors or tensor expressions that result in a tensor invariant of the same kind. A vector cannot be equal to a scalar, nor can a symmetric tensor be equal to an antisymmetric one. Stokes (1845) reasoned that the viscous stress is a symmetric tensor with a zero trace. It can thus only be equal to that invariant of the field of motion that is also a tensor of rank 2, is symmetric, and has a zero trace. Accordingly he reasoned that the viscous stress tensor is proportional to the rate of pure deformation of the fluid motion:

$$\tau_{ik} = \rho \nu d_{ik} = \rho \nu \left[\frac{\partial u_i}{\partial x_k} + \frac{\partial u_k}{\partial x_i} - \frac{2}{3}\frac{\partial u_j}{\partial x_j}\delta_{ik}\right] . \qquad (2.8)$$

The coefficient ν is called the kinematic viscosity, as noted previously. As a consequence of the second law of thermodynamics, its value is never negative.

As a historical footnote, M. Navier in France derived a similar hypothesis using a primitive molecular model about 20 years before Stokes in England. Although the model never achieved acceptance as a basis for explaining physical phenomena, his name is identified along with Stokes as the author of this relationship.

2.3 The Navier–Stokes Equations

We shall now apply the general conservation equation to the three components of momentum, whose magnitudes per unit mass are the velocity components u_i. The sources of momentum in these equations are the sums of the internal and external forces acting on a unit mass. In a nonrotating coordinate system, the internal force consists of gravity whose magnitude is g. In an arbitrary coordinate system we must give it components represented by an index i, but if, as is customary, we define the x_3-direction as the opposite direction of this vector, we can then represent the ith component of this vector by the expression $-\delta_{3i}g$, where δ_{ki} is the unit symmetric tensor defined above.

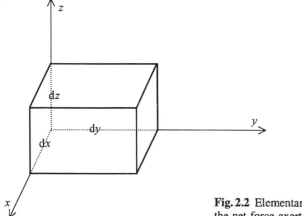

Fig. 2.2 Elementary air parcel for calculating the net force exerted by the stress field

The external forces are those imposed by the stress p_{ki}, consisting of the pressure and viscous forces. We study the net force exerted by all the stresses on the external sides of the surfaces bounding a small parcel, such as that in the preceding figure depicting a parallelepiped of dimensions dx, dy, and dz. For convenience we place the origin of the coordinate system at one corner.

We examine first the net force in the x-direction exerted on the lower, horizontal surface. Since the outer side of this surface is its negative side, the total force exerted on this surface is $-p_{zx}dxdy$, the value of p_{zx} being that for $z = 0$, the assumed height at the lower surface. Next we consider the upper horizontal surface, where the stress has the value $p_{zx} + (\partial p_{zx}/\partial z)\mathrm{d}z$. Since now the outer surface of the volume is the positive side, the total force exerted on the volume is found by multiplying this stress by $\mathrm{d}x$ and $\mathrm{d}y$. When this is done, it is seen that the net force exerted across the two horizontal surfaces in the x-direction is $(\partial p_{zx}/\partial z)\mathrm{d}z\mathrm{d}x\mathrm{d}y$.

We treat the other two pairs of surfaces in like manner, and extend the same analysis to the forces in the other two directions. The result can be written $(\partial p_{ki}/\partial x_k)V$, where V stands for the volume of the parallelepiped, the index i stands for the direction of the force, and k is a dummy index for each of the three pairs of surfaces over which the force components are summed. To get the net force per unit mass, we divide this expression by the density ρ times the volume. This is the source strength that needs to be inserted in the conservation equation for momentum. We see from this analysis the important fact that a simple homogeneous stress field (i.e., a field with identical values everywhere) produces no net force on any portion of the fluid and therefore no source of momentum.

We summarize this discussion by writing the momentum equations resulting when we apply these considerations to the general conservation equation in a nonrotating system:

$$\frac{\partial u_i}{\partial t} + u_k \frac{\partial u_i}{\partial x_k} = -g\delta_{3i} - \frac{1}{\rho}\frac{\partial p}{\partial x_i} + \nu\frac{\partial}{\partial x_k}\left(\frac{\partial u_i}{\partial x_k} + \frac{\partial u_k}{\partial x_i} - \frac{2}{3}\frac{\partial u_j}{\partial x_j}\delta_{ki}\right) \qquad (2.9)$$

This is the general form of the Navier–Stokes equations. The three-dimensional divergence in the last term is negligibly small in large-scale meteorological flows and is usually neglected.

For many purposes in the atmospheric boundary layer, these equations suffice. However, it is sometimes necessary to consider the effects of the Earth's rotation. We shall not derive the Coriolis force that results from this rotation, but will show how this can be represented in tensor notation and added to the Navier–Stokes equations when needed.

As is well known, the Coriolis force can be represented as twice the vector product of the relative velocity V and the Earth's angular velocity vector Ω. When this cross-product is expanded in component form, it becomes

$$C = 2V \times \Omega = 2i(v\Omega_z - w\Omega_y) + 2j(w\Omega_x - u\Omega_z) + 2k(u\Omega_y - v\Omega_x).$$

We can represent the component of C in the ith direction as the sum of nine terms (most of which vanish) involving the unit antisymmetric tensor ε_{ijk}. This tensor has the value zero if any two of the component indices are the same. It has the value 1 if the three indices are cyclical permutations of 123, and -1 if any pair of these indices is reversed. The ith component of the Coriolis force is then given by

$$C_i = 2\varepsilon_{ijk}u_j\Omega_k. \qquad (2.10)$$

2.4 Reynolds Number Similarity

A great deal of design engineering is done by studying the flow of fluids around models on the assumption that the flow around a full-scale structure will be similar to that around the model. The general principle to be followed is that the two flows will be similar if corresponding terms in the Navier–Stokes equations have the same ratios. It is usually not possible to achieve this condition for all the terms, nor is it actually necessary. Many flows are dominated by a limited number of terms, and if the tests can be designed so as to make the ratios of the important terms the same, flow similarity is virtually assured.

The character of flow in pipes is an example. Here, the dominant terms are the nonlinear acceleration terms, often called the inertia terms, and the viscous terms. The ratio of these terms is the Reynolds number which we have already looked at. To see this, we estimate the magnitude of a typical inertia term $u_k \partial u_i / \partial x_k$. If the radius of the pipe is l, and the mean flow speed through the pipe is U, then inertia terms will have an order of magnitude of U^2 / l. By similar reasoning, the viscous terms will have a magnitude of $\nu U / l^2$. The ratio of these terms is the Reynolds number Re which is seen to be

$$Re = \frac{Ul}{\nu} . \tag{2.11}$$

Experiments have shown that when this number is less than about 1000, pipe flow is laminar. While laminar flows can be maintained at higher Reynolds numbers, the flow becomes sensitive to the smallest disturbances, which then grow until the flow becomes turbulent.

2.5 Averaging the Navier–Stokes Equations

Since we have already averaged the general conservation equation, we need only apply that equation to the special case in which q is replaced by u_i, the momentum per unit mass in any one of the three principal directions. We first compare (2.9) and (1.13). The result of carrying out the averaging process on the source term is seen to be

$$\overline{\rho S_{u_i}} = -\overline{\rho} g \delta_{3i} - \frac{\partial \overline{p}}{\partial x_i} + \frac{\partial}{\partial x_k} \left[\overline{\rho} \nu \left(\frac{\partial \overline{u}_i}{\partial x_k} + \frac{\partial \overline{u}_k}{\partial x_i} \right) \right] \tag{2.12}$$

after neglecting the divergence of mean velocity. From (1.19) we obtain the final result of averaging the Navier–Stokes equations;

$$\frac{\partial \overline{u}_i}{\partial t} + \overline{u}_k \frac{\partial \overline{u}_i}{\partial x_k} = -g \delta_{3i} - \frac{1}{\overline{\rho}} \frac{\partial \overline{p}}{\partial x_i} + \frac{1}{\overline{\rho}} \frac{\partial}{\partial x_k} \left[\overline{\rho} \nu \left(\frac{\partial \overline{u}_i}{\partial x_k} + \frac{\partial \overline{u}_k}{\partial x_i} \right) - \overline{\rho u'_k u'_i} \right] . \tag{2.13}$$

We note the fact that each term reduces to the component of a vector in the x_i-direction. As in the general form of the equation, there is a new term not present in the original equation, which can be described as the convergence of an eddy flux. In this case, the property being transported is a vector, so that the flux represents a tensor of rank two. We note further that, like the viscous stress, it is a symmetric tensor. It also appears that the function performed by the negative value of this term, viz. $-\overline{\rho u'_k u'_i}$, is equivalent to the action of a stress. Osborne Reynolds (1895), who first averaged the Navier–Stokes equations in this way used this word to describe the new terms, and today they are known as the Reynolds stress.

The identification of the eddy momentum flux with a force per unit area is more than a convenient analogy. Consider, for example, a uniform mean flow in the x-direction beneath a horizontal surface. Suppose we have a downward flux of x-directed momentum of magnitude $1\,\mathrm{kg\,m\,s^{-1}}$ per unit area per unit time; $\overline{\rho w' u'} = -1\,\mathrm{kg\,m^{-1}\,s^{-2}}$. If there are no forces or other momentum sources, the momentum of the fluid below each unit area of the surface would increase at the rate of $1\,\mathrm{kg\,m\,s^{-1}}$ each second. Now, suppose instead of the flux, we impose a stress in the x-direction equal to 1 Newton per square meter on the top of the surface. Since by Newton's second law, a force is equal to the rate of change of momentum, the resulting change of momentum in the fluid below is the same. For convenience, we shall use the symbol τ_{ki} to denote the total of the two stresses:

$$\tau_{ki} = \overline{\rho} \nu \left(\frac{\partial \overline{u}_i}{\partial x_k} + \frac{\partial \overline{u}_k}{\partial x_i} \right) - \overline{\rho u'_k u'_i} \; . \tag{2.14}$$

It is useful to look at typical orders of magnitude of the two parts of the total stress at the surface layer of the atmosphere, remembering that it is the sum of these two quantities that appears in the last term of the equations of mean motion. We can typically assume that near the ground the mean motion is horizontally homogeneous, i.e., it is independent of x and y, and varies only with height. Because it is nondivergent, there can be no vertical component of velocity if the earth's surface is flat. We define the x-direction as the direction of the wind. Thus only $\overline{u}$ is different from zero and it varies only with height. Using K-theory and (2.13), we can write the sum of the two stresses as

$$\tau_{zx} = \overline{\rho} \, (\nu + K_\mathrm{m}) \frac{\partial \overline{u}}{\partial z} \tag{2.15}$$

in which K_m is the eddy diffusivity for momentum, or, as it is often called, the eddy viscosity. The relative size of the two terms depends only on the relative size of the two viscosities. The molecular viscosity ν for air is about $1.5 \times 10^{-5}\,\mathrm{m^2\,s^{-1}}$. Typical values of the eddy viscosity in the lowest $10\,\mathrm{m}$ of the atmosphere are 1 to $10\,\mathrm{m^2\,s^{-1}}$! Thus the acceleration of the mean motion due to the molecular stress is typically a million times smaller than that produced by the Reynolds stress, and we can safely neglect it in this context.

Since the wind typically increases with height, there is a consistent flow of momentum downward toward the surface. If the surface is smooth, the size of the eddies gets smaller and eventually approaches zero. Exchange theory and experience both show that K_m decreases toward zero, so that there should be a thin layer close to the surface where the molecular viscosity dominates the flow. Another way of looking at this situation is to note that the Reynolds number – formed by putting the length scale equal to the distance from the wall – gets very small. In this kind of environment, the motion becomes laminar, and we call the layer of air close to a smooth boundary the laminar or viscous sublayer. Most natural surfaces are rough, and in such environments, the flow remains turbulent even at the surface.

As we shall show later, whether the surface is smooth or rough, the stress does not get small at the surface; in fact, its value is about the same as it is at a height of ten meters. Thus it is seen that the mean wind shear becomes very large at the surface, and the viscous stress may become dominant over any Reynolds stress that remains. The effect of this stress is a loss of the atmosphere's momentum, an effect we commonly call friction. At a height of a few meters, viscosity has a negligible effect on the mean motion, but the Reynolds stress and its frictional effect are still there in nearly undiminished magnitude.

It should be noted that there exist within the fluid viscous stresses associated with the eddy motions, given by terms like $\rho\nu\partial u'/\partial x_k$. The mean value of such terms is zero, so they have no effect on the mean motion. However, they have a powerful effect on the individual eddies, particularly the smallest ones in which the velocity gradients are largest.

2.6 Problems

1. The kinematic viscosity for water is roughly $10^{-6}\,\mathrm{m^2\,s^{-1}}$. Assuming that the ducts in trees that carry water up from the roots are about one-tenth of a millimeter in radius and that the flow rate is something like $100\,\mathrm{m}$ every 3 hours, compare the Reynolds number for this flow with that in a pipe $1\,\mathrm{cm}$ in diameter at a speed of $10\,\mathrm{m\,s^{-1}}$. Would you expect these flows to be turbulent?

2. Given typical values of the Reynolds stress in the surface layer and a typical wind distribution up to the top of the atmosphere, how long would it take eddy friction to bring the atmosphere's mean momentum to zero (assuming that the stress did not change with time)? Hours, days, weeks, or months?

3. Write the dot product and the cross product of two vectors A and B in tensor notation. Show that $A \times B$ is minus $B \times A$.

4. Show that $\dfrac{1}{3}\dfrac{\partial}{\partial x_k}(p_{jj}\delta_{ki}) = -\dfrac{\partial p}{\partial x_i}$.

5. Vorticity is commonly thought of as a vector, but in tensor notation it is an antisymmetric tensor of rank 2, not rank 1. We commonly are puzzled by this fact and invoke the "right-hand-rule" to make the connection. Discuss the relation between the vorticity tensor and the vorticity vector. Write an equation that relates the two.

6. As pointed out already, the tensor properties of viscous and Reynolds stresses are very similar. However, unlike the viscous stress, the trace of the Reynolds

stress is not zero. Thus the Reynolds stress is more like p_{ki} than it is like the viscous stress. Using the same procedure that was used to define the pressure, separate the Reynolds stress into two parts: an "eddy pressure" and a true "eddy viscous stress." If one combines the two pressures into a "total mean pressure" will the result be greater or less than $\bar{p}$?

7. From the typical vertical distributions of stress and velocity in the air above the ground, how would you expect the eddy viscosity to vary with height in this layer? Can this distribution be justified in the light of the mechanism of eddy viscosity?

8. In its simplest configuration a star is a ball of gas in hydrostatic equilibrium such that the forces of gravity tending to compress the star are balanced by the pressure-gradient forces. Some stars are not in equilibrium, but rather pulsate (i.e., expand and contract rhythmically) about the equilibrium state. If the Stokes equations for the viscous stresses are correct, how would viscosity affect such oscillations?

3 The Neutral Surface Boundary Layer

3.1 Overview of the Atmospheric Boundary Layer

When flying in cloud-free continental areas, it is commonly observed that the air in the lowest one or two kilometers is quite heavily polluted. Visibility over thousands of square kilometers may be restricted to a few kilometers or less. The air seems to be uniformly dirty. Quite surprisingly, when one ascends through the top of this layer, one observes a sharp, distinct top surface. Suddenly one is able to see mountains or clouds hundreds of kilometers in the distance.

The dirty layer is obviously well mixed, for pollutants that must have originated close to the surface are uniformly distributed both vertically and horizontally. For this reason, it is commonly called the mixed layer by the air-pollution community. Meteorologists call it the *atmospheric boundary layer* (ABL), or just as frequently, the *planetary boundary layer* (PBL). These latter names are preferred, because they give recognition to the fact that the main properties are derived from contact with the surface, in contrast to the higher layers above it.

The distinct upper boundary is mainly present in anticyclonic areas where the air is slowly subsiding at rates of the order of $1 \, \mathrm{cm \, s}^{-1}$ (see Fig. 3.1). This rate of subsidence would reduce the depth of the ABL by a few hundred meters per day

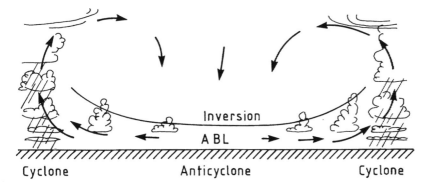

Fig. 3.1 Schematic overview of the atmospheric boundary layer (ABL)

if it were not for the presence of penetrative convection. This process, which over land occurs mainly in the afternoon, and over oceans in areas of cold advection, sends up warm thermals that entrain clean air above down into the ABL and move its boundary upward. The result is a *status quo*, such that the mean thickness of the ABL is determined by the relative strengths of the two processes.

During its contact with the underlying surface, the ABL accumulates its various properties – water vapor, momentum (or its absence), heat, and of course pollutants of many kinds. Through mixing, the ability of the ABL to store these properties is considerable. As time goes on, the horizontal divergence associated with the subsidence gradually carries the accumulated properties outward into the surrounding areas of clouds and precipitation. The ABL gets deeper and less clearly defined. Eventually the air that composed it gets carried aloft, and the products it contained are either rained out or absorbed into the main portion of the atmosphere. Eventually, the cleaned air returns at higher levels to the anticyclonic areas and repeats the cycle.

Of course, the pressure systems in middle latitudes move differently from the airmasses, so that the processes discussed above take place with frequent interruptions, much as a serial television drama. Occasionally, a large anticyclone takes up a temporary residence over an industrially active region in what is called a *stagnation episode*. Some of the worst air-pollution disasters have been associated with these events.

As we look at the mean state of the ABL in horizontally homogeneous areas, we observe that the rate at which any property $\bar{q}$ changes with time is uniquely related to the rate at which the total flux, in this case the eddy flux, of the property changes with height:

$$\frac{\partial \bar{q}}{\partial t} = -\frac{1}{\bar{\rho}} \frac{\partial}{\partial z} \left(\overline{\rho w' q'} \right) . \tag{3.1}$$

Thus we can estimate the mean rate of change of flux with height by looking at the mean rate at which $\bar{q}$ increases with time. We shall consider the details of the flux distribution at a later time. For now we can look at the average rate of decrease of flux with height by assuming that the flux disappears at the top of the ABL. Since the ABL is nominally 1 km thick, we see that the average rate of decrease is about 1% of the surface value for each 10 m increase of height. Actually, close to the surface the rate of change is somewhat larger than it is higher up, but the order of magnitude is correct.

When we look at the mean vertical distribution of properties like wind speed, temperature, or water vapor concentration, we find a different situation. Most of the difference between the surface and the top of the ABL occurs in the lowest 10 m. Compared to the rate at which most things change close to the ground, the flux is for all intents and purposes constant. Higher up, the rates of change with height of the fluxes and other properties are more nearly alike, and the methods needed to understand the distributions are different. For these reasons, it is useful to separate the ABL into two regions. The lowest 10 m are sometimes called the *surface layer, Prandtl layer,* or *constant-flux layer* (an obvious but understandable misnomer), while the remaining ABL above 10 m is called the *outer layer, spiral layer*, or *Ekman layer*.

In the remainder of this chapter, we focus our attention on the vertical distribution of mean quantities in the surface layer. For any property q, the main equation is (1.25) in which it is assumed that the flux is independent of height. The aim is to express the gradient of $\overline{q}$ as a function of z, and then by integration to obtain $\overline{q}$ as function of height. It is clear that the relation between the gradient of $\overline{q}$ and the eddy diffusivity is a reciprocal one. At heights where the exchange is efficient and vigorous, the mean gradients are small and, conversely, where the exchange is inhibited, the gradients must be correspondingly large.

As we have seen, the magnitudes of the exchange coefficients are determined by the size of the energy containing eddies and the magnitudes of their eddying velocities. Both of these quantities are influenced by convective heating as well as other factors. In this chapter we shall simplify the problem by neglecting the effects of surface heating. This, in effect, means that we confine our attention to neutral layers. In the absence of evaporation, which we will assume for now, a neutral layer is one with a an adiabatic mean temperature lapse rate.

3.2 Wind Distribution in the Neutral Surface Layer

Our aim is to try to use our understanding of turbulence to deduce the vertical distribution of mean wind under stationary horizontally homogeneous conditions. For this purpose we look again at (2.15). Without molecular viscosity this becomes

$$\tau_{zx} = \overline{\rho} K_m \frac{\partial \overline{u}}{\partial z} \,. \tag{3.2}$$

We have seen that if we can make a hypothesis about the vertical distribution of the eddy viscosity K_m, the reciprocal relation can lead to a solution. The eddy viscosity K_m has dimensions of velocity multiplied by length, and it is reasonable to try to deduce which of the environmental variables are the correct ones to use. The velocity that is sought should be related to the mean speed of the turbulent eddies. One might suppose that this velocity should be proportional to the wind speed itself, but this is found to be incorrect. The correct estimate of this velocity, which is denoted by u_*, is derived from the stress itself through the equation

$$\tau_{zx} = \overline{\rho} u_*^2 \tag{3.3}$$

which is dimensionally consistent and, like the eddy velocities in neutral conditions, independent of height. This velocity scale is usually called the *friction velocity*. Observations show that the root-mean-square vertical velocity σ_w is independent of height in neutral surface layers and about 25% larger than u_*. Thus u_* serves as a good indicator of the velocity scale required by exchange theory.

The size of the largest eddies, also needed for exchange theory, is provided by the height, since the distance from the ground puts an upper limit on the eddy sizes. Accordingly, we set

$$K_m = ku_* z \tag{3.4}$$

where the constant of proportionality k, known as the *von Karman constant*, has the experimentally determined value of 0.4. With the use of (3.2) and (3.3) we can form a dimensionless wind shear given by

$$\varphi_m \equiv \frac{kz}{u_*} \frac{\partial \overline{u}}{\partial z} = \frac{k}{u_*} \frac{\partial \overline{u}}{\partial \ln z} \tag{3.5}$$

which will prove to be useful later on. This quantity is a function of the heating rate, but for a neutral surface layer it has a value of one.

We see immediately that if we plot measurements of mean wind speed at several heights on a logarithmic scale, they should fall on a straight line with a slope of k/u_*. We can use such observations in either of two useful ways. Assuming that we know the value of k, we can get the value of the surface stress using (3.3). Conversely, we can use independent measurements of the stress to obtain measurements of the von Karman constant k.

Two methods have been used to obtain fundamental measurements of the surface stress. One involves the use of drag plates constructed with sensitive instruments to try to measure directly the drag force exerted on the ground. The problem with such measurements is the difficulty of making certain that the measured drag is representative of the surrounding surfaces. Past attempts to use this method have not given results consistent with other methods. The second method is called the eddy-correlation method and is based on the direct measurement of u' and w' over a sufficiently long period of time to get a representative average of their product. This method requires the use of instruments that can respond very rapidly to the three-dimensional fluctuations of velocity. For many years, small cup anemometers constructed of thin wire and plastic were used along with bivanes that respond both to vertical and horizontal changes of wind direction. More recent methods are based on measuring the speed of sound over short distances (sonic anemometers) and hot-wire anemometers constructed of wires a few micrometers thick. The temperature of such a wire, when heated electrically, is sensitive to air motions and responds almost instantly to their fluctuations.

The von Karman constant has been extensively measured in the laboratory, in wind tunnels, and in the atmosphere. The measured values range from about 0.38 to 0.45. There has been some speculation over the possibility that it depends on the Reynolds number or other aspects of the flow. Due to the uncertainty of its exact value, we use a single significant figure for its value.

The equation for the vertical distribution of mean wind is easily obtained from (3.5) by integration. This is generally written in the form

$$\overline{u} = \frac{u_*}{k} \ln \frac{z}{z_0} \tag{3.6}$$

where the constant that results from the integration, z_0, is seen to be the height at which the mean wind speed calculated by the equation goes to zero. This height or length is called the *roughness length* or *roughness parameter*, and is to some extent indicative of the roughness of the surface. Equation (3.6) was first derived by Prandtl (1932) and is often called the *law of the wall*.

Figure 3.2 illustrates some mean wind measurements made during a large field experiment conducted at O'Neill, Nebraska on the Great Plains of the United States in 1953. The results of this program have been published in two volumes by Lettau and Davidson (1956). The site was selected because it was the flattest and most uniform that could be found. During the course of the measurements, the surface was periodically mowed to keep the height of the grass uniformly a few centimeters tall. Three measured wind distributions are plotted. The first was measured at 1430 hours local time when the stratification was unstable (lapse). The last was observed at 1830 when the stratification was stable. The third, observed at 1630, is approximately neutral when the conditions we have assumed in this chapter prevailed. It can be seen that the observed velocities, when plotted against height on a logarithmic scale, fall nearly on a straight line. By extending this line downward below the heights of observation, we come to a zero mean wind speed at a height of 0.6 cm. This is the roughness length characteristic of O'Neill, Nebraska. As is generally found at most sites it is one seventh to one tenth of the actual height of the grass at the site.

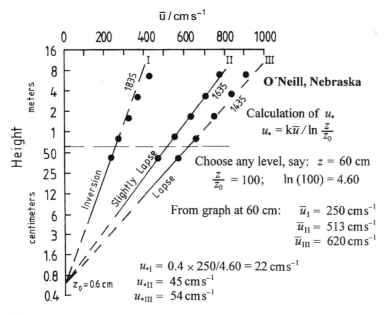

Fig. 3.2 Logarithmic mean wind profiles observed at O'Neill, Nebraska, during the Great Plains Field Experiment in 1953

Since the logarithmic shear is constant, we can evaluate it over the height difference between, say, 60 cm and 0.6 cm so that the speed difference is just the wind speed itself at 60 cm. Reading this from the graph we find a speed of $513\,\text{cm s}^{-1}$. If we take the value of 0.4 for the von Karman constant k, we then find the value of u_* to be $45\,\text{cm s}^{-1}$. The corresponding value of the stress is $0.24\,\text{N m}^{-2}$. Armed with these values we can use (3.6) to calculate the wind at other heights. For example, at a height of 32 m we would expect to observe a mean wind speed of $965\,\text{cm s}^{-1}$.

So far we have only justified (3.6) on the basis of conditions that we know to exist in the lowest 10 meters of the atmosphere, and it could be asked why we apply the result at a height of 32 m. The answer is that experience shows that the logarithmic distribution gives good results up to heights of at least 100 m, even though the assumptions we have made in this section begin to fail quite badly. It should be noted, and we shall see the reason for this later, that the sensitivity to stratification becomes greater the higher one goes. If we stay close enough to the ground, even the non-neutral profiles at 1430 and 1830 are quite well described by (3.6), even though at higher levels the departures are conspicuous. Thus, the question of how high we can trust the profile to be logarithmic depends mostly on how sure we can be that the stratification is exactly neutral.

Once z_0 has been determined, it is possible to obtain u_* from a single observation of the wind, at least in principle. The value of z_0 depends only on the geometry of the surface and is not generally dependent on the value of u_* or the stability. Exceptions occur, such as, for example, long grass and water bodies, the surfaces of which are geometrically disturbed by the wind. Also, experience shows that the roughness that counts is that which is representative of the surface in the upwind direction, out to a distance of 100 times the height of the topmost wind measurement. At some sites the effective roughness varies with different wind directions. Typical roughness lengths found in various types of terrain are given in Table 3.1.

Table 3.1 Typical values of the roughness length z_0 and drag coefficient C_d for various surfaces

Type of surface	z_0/m	C_d $(z = 10\,\text{m})$
Mud flats or smooth ice	1.0×10^{-5}	0.0008
Smooth snow	5×10^{-5}	0.0011
Grass up to 1 cm	0.001	0.002
Grass up to 60 cm	0.05	0.006
Pasture land	0.20	0.010
Suburban landscapes	0.6	0.02
Forests and cities	1 to 5	0.03 to 0.33

Over water surfaces, the shape of the surface and the roughness lengths vary with the wind speed. In general, the ocean surface is remarkably smooth, with values of 10^{-4} to 10^{-5} m being typically observed from neutral wind profiles.

The smallness of these roughness length values seems to bear little relationship to the much greater height of the waves. Munk has pointed out that the drag of waves and other surface objects is highly dependent on their shape. Probably only the steepest of these are effective. Thus the roughness length probably depends only on the short capillary waves and very little on the longer waves that give the greatest visual impression of roughness. A second point, often lost sight of, is that the roughness elements on the surface are not stationary as they are over land. If one transforms the coordinate system so as to move along with the significant waves on the surface and effectively make them stationary, it is found that the profiles are also logarithmic, but the corresponding roughness lengths are considerably larger. (See Problem 3 at the end of this chapter.)

With very light wind speeds, the ocean surface is smooth. In this case dimensional considerations suggest that z_0 should be proportional to ν/u_*. Over such surfaces, which may also include flow over smooth ice, the effective roughness decreases with wind speed at a fixed height. Roll, using observations of Motzfeld, concluded that z_0 decreases with increasing u_*, a fact that might be explained by a tendency for larger waves – present with higher wind speeds – to shelter portions of the sea surface from exerting a drag on the wind. However, later data from a variety of sources tend to show a rapid growth of z_0 with increasing values of u_*. Charnock, reasoning that the roughness should depend only on the surface stress of the wind and gravity, has fitted his and other observations to the dimensionally derived formula

$$z_0 = 0.0123 u_*^2/g \; . \tag{3.7}$$

This formula is widely quoted at the present time. However, it has been extensively criticized. Efforts to improve on it have been undertaken in recent years without much success.

Von Karman has discussed the question of how to differentiate aerodynamically smooth surfaces from those that are rough. The question can be logically discussed by looking at the Reynolds number of the surface elements comprising a length h_0 – the average height of the surface obstacles – a velocity u_* formed from the surface stress, and the viscosity ν. When this number is less than one, the surface obstacles are entirely contained within the laminar sublayer and unable to generate significant turbulence. If, on the contrary, h_0 is large compared to ν/u_*, turbulence extends all the way to the surface, and the laminar sublayer effectively ceases to exist.

On aerodynamically smooth surfaces the stress is exerted by viscosity and at any fixed height within the laminar sublayer, the magnitude of the stress is proportional to the wind speed. On rough surfaces, the effects of viscosity are negligible, and the transfer of stress to the underlying surface is accomplished by form drag, i.e., pressure differences between the upwind and downwind sides of the obstacles. This kind of drag is proportional to the square of the mean wind speed and tends to depend on the geometrical form of roughness elements as well as their spacing and arrangement. Thus, one cannot judge the value of z_0 from the height of the obstacles alone.

3.3 Mean Flow in the Vicinity of the Surface

In the immediate vicinity of the surface, the flow is controlled strongly by the surface roughness, and it is insensitive to the effects of stratification. Thus (3.6) is quite broadly applicable in this region. We must be careful, however, not to apply this equation to the flow at heights that are closer to the surface than the highest obstacles, and possibly somewhat higher. The reason is that this equation depends on the constancy of the stress transferred by the turbulence. As one drops below the height of the obstacles, an increasing portion of the stress is transferred by the obstacles, so that although the total stress is still independent of height, the part transferred by the turbulence is no longer constant. Although the roughness length is defined by the mean wind equation, it has no significance with respect to the real wind distribution close to the surface.

The closer one gets to the surface, the less easy it is to define where the surface actually is, and the more important it becomes to define the dynamically significant zero point of the height scale. In the case of a dense forest canopy, for instance, the wind above may have no way of knowing where the actual ground is, nor does it care. It is the configuration of the treetops and perhaps the density of the vegetation that determines the aerodynamic zero point. The difference between this aerodynamic zero point and the zero point of the scale used for the actual height measurements is called the *zero-plane displacement*. We shall denote this length by D. We apply it by subtracting it from the actual measured height in order to obtain the dynamically effective height. The wind equation then is

$$\bar{u} = \frac{u_*}{k} \ln \frac{z - D}{z_0} \ .\tag{3.8}$$

The value of D is generally obtained by fitting low-level observed profiles to (3.8); the procedure for neutral conditions is to try different values of D until one is found that puts the measured points on a straight logarithmic line. This correction is important for the wind speed within a few D units of the top of the vegetation and becomes negligible when the height is large compared to D. Like z_0, D is independent of stratification or speed in most environments. This procedure can also be used in non-neutral conditions if the lowest levels are close enough to the surface, because the effects of buoyancy are minimal at these levels.

3.4 Miscellaneous Topics

Climatologists and others frequently look for short-cut methods to estimate the drag of the wind on the surface from measurements of wind speed. A formula that is commonly used is

$$\tau_{zx} = C_d \, \bar{\rho} \, \bar{u}^2\tag{3.9}$$

in which the coefficient C_d is known as the *drag coefficient*. From a comparison of this equation with (3.3) and (3.5), we see that as long as the winds are observed at a fixed height z, the drag coefficient in neutral surface layers is independent of the wind speed, and the surface stress is indeed proportional to the square of the wind speed. We must remember, however, that the relation between wind speed and surface stress depends on the height at which the wind is measured, and that there is a great deal of variability in these heights, over both land and ocean. In addition, we shall see later that the requirements for horizontal homogeneity that are implicit in these equations is so difficult to achieve, that very few sites can qualify. Measurements made aboard ships are especially subject to failures of this requirement because of local disturbances of the wind flow.

Engineers are fond of power laws, and have traditionally estimated winds at heights other than those for which they have observations by the use of the equation

$$\frac{\overline{u}}{u_1} = \left(\frac{z}{z_1}\right)^p \tag{3.10}$$

in which u_1 is the known mean wind speed at a height z_1, and $\overline{u}$ is the desired wind speed at some other height. The exponent p has generally been assumed to be 1/7, and (3.10) is often called the *one-seventh power law*.

To understand the meaning of this equation, consider Fig. 3.3. Let us suppose that p in (3.10) has the value 0. In this hypothetical case, $\overline{u}$ would equal u_1 at every height except $z = 0$, where its value would be indeterminate. Alternatively, a value of $p = 1$ would cause the wind to vary linearly from zero at the surface to u_1 at z_1. In order to approximate the profile of Prandtl's law of the wall,

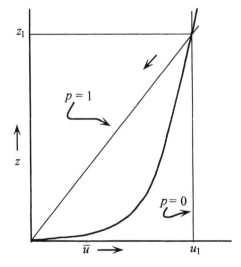

Fig. 3.3 Power-law wind profiles. The value of the exponent p is sensitive to the roughness

p must lie somewhere between 0 and 1. The value of p is not constant but depends on surface roughness. Increasing the surface roughness increases the relative efficiency of mixing at the lower boundary. The reciprocal relation discussed above then indicates that the velocity gradient must then diminish close to the surface. The profile in this case must change so that p becomes greater. The value of p is also dependent on the stratification.

3.5 Distribution of Passive Mean Properties

Most properties other than momentum and heat are passive in the sense that their distribution plays no part in the turbulent mechanisms that spread them. Included in the broad list of such properties are most contaminants and water vapor, to the extent that we can overlook its small effect on the density and buoyancy of air particles. We can quite easily derive the distribution of these quantities by following the same procedure we have carried out for momentum.

In place of u, we consider the more general variable q, which can stand for the amount of any passive quantity per unit mass. The analogue of (3.2) is

$$-\overline{\rho w'q'} = \overline{\rho} K_q \frac{\partial \overline{q}}{\partial z} \tag{3.11}$$

in which K_q stands for the eddy diffusivity of the particular property. The presence of a minus sign in contrast to (3.2) comes about because the stress is by definition minus the flux. Following the former procedure again, we define a new parameter q_* by the equation

$$-\overline{\rho w'q'} = \overline{\rho} u_* q_* \tag{3.12}$$

which is analogous to (3.3). It should be noted that q_* has the same sign as the gradient of $\overline{q}$ – negative for an upward flux of q and positive if downward. Elimination of the flux itself between (3.11) and (3.12) and u_* using (3.4) provides an equation for the dimensionless gradient of $\overline{q}$.

$$\varphi_q \equiv \frac{kz}{q_*} \frac{\partial \overline{q}}{\partial z} = \frac{k}{q_*} \frac{\partial \overline{q}}{\partial \ln z} = \frac{K_m}{K_q}, \approx 1 . \tag{3.13}$$

We are now confronted with one of the most important questions of atmospheric turbulence – the ratio of the exchange coefficients for passive scalar properties and for momentum. This ratio is a function of the stability, but for a neutral boundary layer we can consider it to be constant. Not all experimental data are consistent in showing what this constant is. Until this question is clarified, it is probably best to assume a value of unity. This assumption is sometimes referred to as the *Reynolds analogy*.

Thus we see again that if observations of $\overline{q}$ observed simultaneously at several heights are plotted against height on a logarithmic scale, they fall on a straight line with a slope of k/q_*. Thus, with the aid of the assumption of Reynolds' analogy, we can use observations of the distribution of $\overline{u}$ and $\overline{q}$ in the surface layer to estimate the flux of the property q.

Unlike the momentum, the mean value of a passive property does not usually approach zero at the surface, but rather some value that depends on processes going on at the boundary or within the medium that underlies it. The kinds of interaction that determine this value probably differ for each property and probably for each kind of underlying medium. For now, we can only give it a symbol, say, q_a. Integration of (3.13) then leads to the equation

$$\overline{q} - q_a = \frac{q_*}{k} \ln \frac{z}{z_a} \, . \tag{3.14}$$

The constant of integration z_a is the height at which $\overline{q}$ is equal to the surface value q_a. It is the analogue of the roughness length, and often it is assumed to be the same. Considering that the mechanisms that produce z_0 and z_a are completely different, this assumption is almost certainly incorrect. Unfortunately there is little guidance to be had from observations.

We can also find an equation analogous to the drag coefficient (3.9) by eliminating q_* using (3.12) and (3.6). This equation may be written

$$-\overline{\rho w'q'} = C_q \overline{\rho} \, \overline{u} \, (\overline{q} - q_a) \tag{3.15}$$

where the bulk transfer coefficient has the value

$$C_q = \frac{k^2}{\left(\ln \frac{z}{z_0} \ln \frac{z}{z_a} \right)} \, . \tag{3.16}$$

It is assumed here that the wind and the value of $\overline{q}$ refer to the same height z; otherwise, appropriately different heights should be used in the denominator. Equations such as (3.15) are frequently used as a basis for estimating the evaporation rate from the ocean and from lakes.

In the air pollution community, the downward flux of a pollutant close to the surface is called the deposition rate. This quantity has received a great deal of attention, since most models for predicting mean concentration distributions must state this quantity as a boundary condition. The basis usually used for this purpose is the so-called *deposition velocity* V_d defined by the equation

$$-\overline{\rho w'q'} = \overline{\rho} V_d \overline{q} \, . \tag{3.17}$$

The only good thing that can be said about this equation is that it is dimensionally correct. Since it is physically incompatible with (3.15) there is no logical reason to think that the deposition velocity should be related to the wind speed or other purely meteorological parameter.

3.6 Problems

1. Assume a mean evaporation rate at the surface equal to the global average of about 100 cm of liquid water per year, and that this flux decreases linearly to the top of the ABL at a height of 1 km. If we neglect the variation of density in this layer, what is the mean rate of decrease of the water vapor flux with height? By what percentage does the flux change in the lowest 10 m?

2. In the above example, what is the water-vapor flux at the surface in SI units, and by what amount does the average water vapor concentration, expressed in parts per thousand (‰) change each 24 hours, assuming horizontal homogeneity? Obviously the climatic average rate of change is close to zero. How can this be explained?

3. Over land the obstacles that create surface form drag are stationary, but over the ocean they tend to move at a finite speed in the direction of the wind. Make a copy of the neutral wind profile observed at 1635 local time in Fig. 3.2. Go through the analysis routine with the assumption that z_0 is the level where the wind speed is equal to the speed c of the travelling surface obstacles. Assume that c is greater than zero. What conclusions can be drawn from this excercise about the effect of wave movement on the drag coefficient and on the ratio of z_0 to the height of the waves?

4. Consider the surface layers of the atmosphere and the ocean that are in contact with each other. What is the typical value of the *ratio* of the stress within the atmosphere to that within the ocean surface layers? What is the ratio of the values of u_*?

5. Find an equation that relates the exponent p in the power law with the roughness length of the surface. (This relation will include the heights z and z_1.) Show that when z and z_1 are kept fixed, the value of p increases as z_0 increases.

6. Prandtl developed a theory of flow in the neutral surface layer by postulating that K_m is proportional to the square of the height and to the vertical gradient of the mean wind. Examine this hypothesis for dimensional validity and then find the resulting mean wind profile.

7. On another occasion Prandtl developed a theory of the surface layer wind profile assuming that the mixing length l, is proportional to the ratio of the first and second derivatives of the mean wind with respect to height. Derive the wind profile that satisfies this hypothesis.

8. Taylor (1932) made a first-order closure for the Reynolds stress by assuming that the vorticity $(\partial w/\partial x) - (\partial u/\partial z)$, rather than the momentum u, is conservative. He began with the equation

$$\frac{\partial \left(-\overline{u'w'} \right)}{\partial z} \equiv -\overline{u'\frac{\partial w'}{\partial z}} - \overline{w'\frac{\partial u'}{\partial z}} \tag{3.18}$$

and assumed that $\overline{u'^2}$ and $\overline{w'^2}$ are independent of x and that the turbulence is two-dimensional. In this way, derive Taylor's result:

$$\frac{1}{\rho}\frac{\partial T_{zx}}{\partial z} = K_m \frac{\partial^2 \overline{u}}{\partial z^2} \; . \tag{3.19}$$

How does this result differ from (3.2)? Which equation compares better with observations?

4 The Energy Equations of Turbulence

The single most useful measure of the intensity of turbulence is the *turbulent kinetic energy* (TKE) per unit mass, defined as $\frac{1}{2}\overline{u_i'^2} \equiv \frac{1}{2}\overline{u_i'u_i'} \equiv \frac{1}{2}\left(\overline{u'^2 + v'^2 + w'^2}\right)$.

In this chapter we shall look at the equations that describe three kinds of atmospheric energy: the energy of mean motion, the TKE, and the thermodynamic internal energy $c_v T$.

We have seen from exchange theory that eddy diffusivity estimates require us to know a representative value of the turbulent velocity. Such a value is easily obtained from the TKE if it is known. We were able to estimate K_m in neutral surface layers by the use of u_* alone, but when heating is taking place, it is necessary to consider the full TKE equation. The TKE equation is also needed in order to address the question of the criterion for the existence of turbulence in stable boundary layers.

Before discussing the equations of turbulent energy, we must first review the energy equations of the instantaneous state of a fluid. In this chapter we shall develop the subject somewhat heuristically emphasizing concepts rather than rigor. In Appendix A a more rigorous treatment will be found. We shall emphasize in both treatments that it is possible to arrive at a comprehensive understanding of energy transformations in a moving fluid without making the restrictive assumptions that accompany the classical treatment of thermodynamics.

4.1 Energy of the Instantaneous State of a Fluid

We begin with a statement of the conservation of total energy. Specifically we can state this principle by stipulating that the change of energy of any closed system cannot arise out of processes occurring inside the system. It must therefore be equal to the total rate work is done on its boundary plus the net heat entering the system through its boundary.

Energy is a function of the instantaneous state of the system, which in turn is determined by its temperature, volume (in some cases), position of its center of mass, velocity, and possibly other things such as its composition. The function that defines energy is not *a priori* evident (though it may seem so to those who have been educated in such matters). The founders of thermodynamics had to determine it experimentally, and it is useful to ask how this can be done operationally. In principle one proceeds as follows.

We first insulate the closed system so as to ensure that no heat enters or leaves it. We then do a measured amount of work on it and observe the changes in its variables of state. It has been found through numerous such experiments, that there always exists a functional relation between the two and what that function is for any given system. Thus an evaluation of the rate at which work is done on a unit mass of air is fundamental in arriving at the total energy equation. We shall assume the absence of things like paddle wheels and therefore that all work is done on the system boundary.

4.2 Work Done on the Boundary

We look at two examples that make clear the relationships involved in this process. Consider the simple configuration of a section of a pipe of 1 meter square cross-section, confined by moving partitions (pistons), see Fig. 4.1. The pressures and velocities characteristic of the fluid at each end are p_1, u_1, and p_2, u_2. Since the walls of the square tube do not move, no work is done on their surfaces. Therefore, the rate at which work is done on the boundary of the air system is

$$(p_1 u_1 - p_2 u_2)(1m^2) = \frac{-(p_2 u_2 - p_1 u_1)}{x_2 - x_1} V \ . \tag{4.1}$$

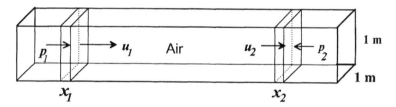

Fig. 4.1 A thermodynamic system that changes at a finite rate

We shall normalize this result by dividing by the mass, ρV. The result can be expressed more generally in the form of derivatives

$$W_p = -\frac{1}{\rho}\frac{\partial}{\partial x}(pu) \equiv -\frac{p}{\rho}\frac{\partial u}{\partial x} - \frac{u}{\rho}\frac{\partial p}{\partial x} = W_{p1} + W_{p2} \ . \tag{4.2}$$

The first term W_{p1} consists of minus the pressure times the divergence rate. The divergence rate can be defined as the fractional rate of change of the volume of a parcel if the parcel has a unit mass. Thus we can write

$$W_{p1} = -p\frac{d\alpha}{dt} \tag{4.3}$$

in which α stands for the specific volume (volume per unit mass). W_{p1} is just the familiar work function that appears in the limited statement of the first law of ordinary thermodynamics, except that we have made no restrictions on the rate at which this work is done.

There is, however, another part of the work done on the boundary that is not ordinarily considered in thermodynamics, namely, W_{p2}. This part is seen to be the rate at which work is done by the pressure-gradient force in the Navier–Stokes equations. This force plays a role in the acceleration of air, and may change the kinetic energy of the fluid in the process.

Consider now another example of fluid confined between two horizontal plates moving at different speeds in the x-direction (Fig. 4.2). The rate at which work is done on the fluid consists of that done by the forces acting on the top side of the top plate plus that done on the under side of the bottom plate. For each unit of horizontal area, the rate at which work is done on the top (subscript t) surface is $\tau_t u_t$, where τ_t is the component of the stress acting horizontally on the top surface in the direction of motion. In like manner, we see that the rate at which work is done on the under side of the lower surface is $-\tau_b u_b$ per unit area. We must add these together, and divide by the mass of confined fluid per unit horizontal area to get the result

$$W_v = \frac{1}{\rho} \frac{(\tau_t u_t - \tau_b u_b)}{\delta} \approx \frac{1}{\rho} \frac{\partial (\tau u)}{\partial z} \,. \tag{4.4}$$

As before, we can divide this rate of work into two parts:

$$W_v = \frac{\tau}{\rho} \frac{\partial u}{\partial z} + \frac{u}{\rho} \frac{\partial \tau}{\partial z} = \varepsilon_T + W_{v2} \,. \tag{4.5}$$

Looking at the second one first, we see the rate at which work is done per unit mass by the stress-gradient force in the Navier–Stokes equation. Since the pressure does no work on these horizontal surfaces, we are looking at work done entirely by the viscous stresses. This part of the work evidently affects the kinetic energy of the fluid through the accelerations accomplished by viscous forces.

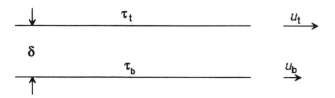

Fig. 4.2 Fluid moving in a shearing configuration between two parallel plane boundaries

Next we look at the first term in (4.5) which is denoted by ε_T. It is the counterpart of the pressure expansion term of the first law, but it is accomplished by viscosity. In this example the stress is proportional to the gradient of the velocity, and we have

$$\varepsilon_T = \nu \left(\frac{\partial u}{\partial z} \right)^2 \tag{4.6}$$

which is seen to be a positive definite form. Unlike the reversible work of expansion, this term can never be negative. It represents the rate of dissipation of mechanical energy into thermal energy per unit mass. We should note that the rate at which work is done by frictional forces in the Navier–Stokes equations may be either positive or negative. Friction is just as important for increasing the velocity as it is for decreasing it. Therefore it is important not to confuse the functions performed by each of these terms.

So far we have looked only at examples. We can generalize these results by stating that in an insulated system the total energy E_T per unit mass changes at the rate

$$\frac{dE_T}{dt} = \frac{1}{\rho} \frac{\partial (p_{ki} u_i)}{\partial x_k} . \tag{4.7}$$

4.3 Heat

In thermodynamics, the word *energy* is reserved for use in representing the state of a system, while *heat* is reserved for use in describing the passage of this property from one system to another. Zemansky has likened *energy* to the water that resides in a lake and *heat* to the rain that may fall into it and thus cause the water in the lake to increase. Heat is energy in transit from one system to another. We should avoid the common mistake of using *heat* as a synonym for internal *energy*.

It will serve our purpose best to define the rate at which heat is transported by using a flux vector h_k. Specifically h_k is the amount of energy transferred by conduction or radiation per unit time and per unit area through a surface normal to x_k. It is positive when its direction of passage is in the positive direction of x_k.

Consider a finite closed system defined by a set of particles that makes up its surface, and let h_n be the component of the heat flux vector normal to this surface, and positive when directed outwardly. The net rate of heating of the system is then

$$\frac{\delta Q}{\delta t} = - \iint_S h_n dS \tag{4.8}$$

where the integration is over the entire surface S of the system. By the use of Gauss's divergence theorem, we can rewrite this as

$$\frac{\delta Q}{\delta t} = - \iiint_V \frac{\partial h_k}{\partial x_k} dV . \tag{4.9}$$

Thus we see that for a small increment of the fluid, the external heating rate per unit volume is the rate of convergence of the heat flux.

We can now remove the restriction we previously placed on the rate of change of total energy in (4.7). In place of this we now have

$$\frac{\mathrm{d}E_{\mathrm{T}}}{\mathrm{d}t} = \frac{1}{\rho} \frac{\partial (p_{ki} u_i)}{\partial x_k} - \frac{1}{\rho} \frac{\partial h_k}{\partial x_k} . \tag{4.10}$$

4.4 The Energy Equations and Energy Transformations

Perhaps rather arbitrarily, we shall assume that the total energy of the atmospheric systems we deal with is composed of kinetic energy of motion, potential energy Φ of the mass distribution, and the thermodynamic internal energy. These energies per unit mass of fluid are given by $u_i^2/2$, gz_p, and $c_v T$, respectively, where z_p refers to the height above an arbitrary reference level of the parcel in question, and c_v is the specific heat at constant volume. Thus (4.10) becomes

$$\frac{\mathrm{d}}{\mathrm{d}t} \left(\frac{u_i^2}{2} + \Phi + c_v T \right) = \frac{1}{\rho} \frac{\partial (p_{ki} u_i)}{\partial x_k} - \frac{1}{\rho} \frac{\partial h_k}{\partial x_k} . \tag{4.11}$$

We can also derive equations for the individual forms of energy. To get an equation for the rate of change of kinetic energy, multiply the Navier–Stokes equation through by u_i, noting

$$u_i \frac{\mathrm{d}u_i}{\mathrm{d}t} \equiv \frac{\mathrm{d}}{\mathrm{d}t} \frac{u_i^2}{2} \tag{4.12}$$

with the result (see (A.14) in Appendix A) that

$$\frac{\mathrm{d}u_i^2/2}{\mathrm{d}t} = -gw + p \frac{\mathrm{d}\alpha}{\mathrm{d}t} - \varepsilon_{\mathrm{T}} + \frac{1}{\rho} \frac{\partial}{\partial x_k} (p_{ki} u_i) . \tag{4.13}$$

The equation for the change of potential energy is simply

$$\frac{\mathrm{d}\Phi}{\mathrm{d}t} = gw . \tag{4.14}$$

Finally, we get the thermodynamic energy equation by subtracting both (4.13) and (4.14) term-by-term from the total energy equation (4.11) to get

$$\frac{\mathrm{d}\,c_v T}{\mathrm{d}t} = -p \frac{\mathrm{d}\alpha}{\mathrm{d}t} + \varepsilon_{\mathrm{T}} - \frac{1}{\rho} \frac{\partial h_k}{\partial x_k} . \tag{4.15}$$

Taken together, (4.13), (4.14), (4.15), and their sum (4.11) give us a comprehensive description of energy changes within a fluid system. We may view the work done by the stresses on the boundary as being the external source of mechanical energy to the system. Part of this energy may be transformed to or from potential energy through the term gw, which appears with opposite signs in (4.13) and (4.14). Likewise, the terms $p(\mathrm{d}\alpha/\mathrm{d}t)$ and ε_{T} represent transformations to and from mechanical and internal energies, since they occur with opposite signs in (4.13) and (4.15). We note also that none of these three transformation terms appears in the total energy equation. Thus they are simply changes of one form of energy into another form.

Equation (4.15) is a restatement of the ordinary first law of thermodynamics, but it differs from the usual statement in that it assumes that processes go on at a finite rate, and it contains in addition the dissipation rate, which is zero when only infinitesimal rates of change are allowed.

We digress for a moment to consider the form of this equation that applies to a continuous medium that is not moving but has a continuously changing temperature in space and time:

$$\frac{\partial T}{\partial t} = -\frac{1}{\rho c}\frac{\partial h_k}{\partial x_k} . \tag{4.16}$$

In such systems, the heat flux is determined entirely by the temperature distribution through the Fourier heat conduction equation,

$$h_k = -\lambda \frac{\partial T}{\partial x_k} . \tag{4.17}$$

Combining the two we get the classical differential equation for the rate of change of temperature at a fixed point:

$$\frac{\partial T}{\partial t} = \frac{\lambda}{\rho c}\frac{\partial^2 T}{\partial x_k^2} = \kappa \frac{\partial^2 T}{\partial x_k^2} . \tag{4.18}$$

In this equation c is the specific heat of the medium (assumed to be incompressible) and κ is the thermal conductivity.

4.5 The Second Law of Thermodynamics

Equation (4.15) may be rewritten in the form

$$\frac{\mathrm{d}\, c_{\mathrm{p}}T}{\mathrm{d}t} - \alpha \frac{\mathrm{d}p}{\mathrm{d}t} = \varepsilon_{\mathrm{T}} - \frac{1}{\rho}\frac{\partial h_k}{\partial x_k} \tag{4.19}$$

by using the equation of state. The left side becomes the time rate of change of a function of state if we divide through by the absolute temperature T. This function is the entropy per unit mass

$$s \equiv c_{\mathrm{p}} \ln T - R \ln p + \text{const} . \tag{4.20}$$

Substituting this in the equation and making some expansions gives the result

$$\rho \frac{\mathrm{d}s}{\mathrm{d}t} = \frac{\rho \varepsilon_{\mathrm{T}}}{T} - \frac{1}{T}\frac{\partial h_k}{\partial x_k} \equiv \frac{\rho \varepsilon_{\mathrm{T}}}{T} - \frac{h_k}{T^2}\frac{\partial T}{\partial x_k} - \frac{\partial}{\partial x_k}\left(\frac{h_k}{T}\right) . \tag{4.21}$$

It is seen that if we define the new vector h_k/T as the flux of entropy associated with the external heat sources, the last term represents the amount of entropy entering a system through its boundary per unit time and per unit volume. This may have either sign depending on the distribution of the heat flux and absolute temperature over the boundary of the volume.

The remaining terms on the right side of (4.21) must represent sources of entropy from inside the volume. In the case of total energy, we have seen that there are no such energy sources. In the case of entropy as we have defined it, the sources consist of the dissipation rate divided by the mean temperature and a term that depends on the relation between the heat flux and the temperature gradient. For molecular conduction in a homogeneous medium we have the Fourier heat conduction (4.17). When this is used to evaluate the second term of (4.21) the result is a positive definite form. Thus we see that as long as the kinematic viscosity and the thermal conductivity are positive, the internal sources of entropy are also positive. The second law relates only to the internal source strengths of entropy, and it states simply that these may never be negative.

4.6 The Boussinesq Approximation

Up until this point in our averaging of the equations, we have ignored density fluctuations. Even though we have allowed, at least implicitly, the possibility of slow variations of mean density, we have assumed that the instantaneous density and mean density are the same. There comes a point, however, when an important question arises. The occurrence of temperature fluctuations in a fluid (and to a smaller extent fluctuations of water vapor concentration) obviously are associated with buoyancy fluctuations that play an important role in the convective production of kinetic energy.

The rigorous inclusion of density fluctuations in all we have done so far would greatly complicate our equations and make it more difficult to understand the important processes that go on. The Boussinesq approximation allows us to neglect the effects of density fluctuations in all the averaged equations except those in which buoyancy plays an important part. Such terms are easily recognized, for buoyancy involves gravity as an essential mechanism. Thus we need only be concerned with density fluctuations when we deal with the transformations between potential and internal energy.

When we look at the causes of density fluctuations, we must examine both temperature and pressure fluctuations. The governing equation is the equation of state

$$p = R\rho T \ .\tag{4.22}$$

Using the Boussinesq approximation, we can easily average this equation

$$\overline{p} = R\overline{\rho T} \ .\tag{4.23}$$

Now for all of the variables, the fluctuations are a small fraction of the mean values, and to a high degree of accuracy we can write

$$\frac{p'}{\overline{p}} = \frac{\rho'}{\overline{\rho}} + \frac{T'}{\overline{T}}. \tag{4.24}$$

In principle, density fluctuations are caused both by temperature fluctuations and also by pressure fluctuations, but in fact the influence of pressure fluctuations are nearly negligible for natural processes that occur in the atmosphere.

Pressure fluctuations that are not balanced by the accelerations associated with turbulent motions are dispersed with the speed of sound and so are quickly attenuated. A good estimate of the pressure fluctuations that are likely to remain can be obtained by subtracting the averaged Navier–Stokes equations (2.13) from the unaveraged ones (2.9). The pressure fluctuations we are concerned with are those associated with the inertia terms, and these have a magnitude of the order of Uu', where U is representative of the mean velocity and u' is representative of the velocity fluctuations. Thus, in view of (4.23) and the well known equation for the velocity of sound, we have

$$\left| \frac{p'}{\overline{p}} \right| \sim \left| \frac{p'}{R\overline{\rho}\overline{T}} \right| \sim \left| \frac{Uu'}{R\overline{T}} \right| \sim \left| \frac{Uu'}{c^2} \right| \tag{4.25}$$

U is typically of the order of $10~\mathrm{m\,s^{-1}}$, and u' is of order $1~\mathrm{m~s^{-1}}$, while the velocity of sound c is about $300~\mathrm{m\,s^{-1}}$. Thus $p'/\overline{p}$ should be of order 10^{-4}. Compared to this, T' is of order 1 K while $\overline{T}$ is or order 300 K; thus $T'/\overline{T}$ and consequently $\rho'/\overline{\rho}$ are one to two orders of magnitude greater than $p'/\overline{p}$. We can use the temperature fluctuations alone to make good estimates of the density fluctuations.

Insofar as the discussion has been based on (4.22) to (4.24), the possibility has to be considered that the gas constant R also might fluctuate as a result of a change of constitution. This possibility becomes a reality when moisture fluctuations occur. Fortunately this problem is easily circumvented by replacing the temperature in these equations by the virtual temperature T_v. To a first approximation, the fluctuation of virtual temperature can be determined from

$$T'_v \approx T' + 0.61\overline{T}q' \approx T' + 180q' \tag{4.26}$$

in which q refers to the specific humidity (i.e. water vapor concentration by mass).

4.7 Open Systems

It has been noted earlier that there are two fundamentally different kinds of thermodynamic systems. Classical thermodynamics, as it is normally presented in meteorological texts, is concerned with closed systems: one follows a volume of the substance that never changes. Following a body of fluid becomes impossible when only the mean motion can be measured. If we insist on trying to follow a mean parcel, or we try to evaluate changes in a coordinate system that moves with the mean velocity, we must recognize that turbulent motions are constantly causing some air to leave our system to be replaced by other air. Since it is impossible to keep track of all the masses that made up the original system, we must revert to defining the system by prescribing the position of its boundaries while allowing the passage of fluid into or out of the geometrically defined volume.

We deal with such systems by keeping a budget of the properties carried across the boundary, much as customs authorities monitor the passage of goods in and out of a country at its borders. By ordinary bookkeeping methods it is a simple matter to calculate the amount of increase or decrease of such articles inside the country as a result of trading with the outside. For the most part this procedure is straightforward, but in the case of heat, special precautions are necessary.

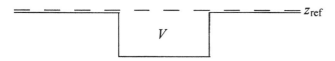

Fig. 4.3 Passage of a volume V of air into an open system comprising the air above the level marked by the dashed line. The solid line traces the initial location of the boundary of a closed system that coincides with the open system at the end of the transfer

An example will make clear the nature of the problem. We define an open system as all the air above some reference level z_{ref}. In Fig. 4.3 this boundary is indicated by a dashed line. We shall evaluate the increase of the internal energy of this system that results from the passage of a volume V of air into this open system from below. For this purpose we define an equivalent closed system, bounded by the solid line, containing all the air that will reside within the open system at the end of the transfer. We note that two things happen during the transfer: (1) the entering air adds its internal energy to what was initially contained in the open system; and (2) work is done on the boundary of the closed system in making the transfer. The required bookkeeping entries are the following:

Energy transported by the mass ... $= c_v \rho T V$
Work done on the closed system ... $= pV$
Energy increase of the open system
   resulting from the transfer ... $= (c_v \rho T + p)V = c_p \rho T V$

The result is that the energy increase is not the energy of the air that is brought in, but rather its enthalpy. We thus have the rule that the heat brought across the boundary of an open system is equal to the enthalpy of the transporting mass.

We can now proceed to average the first law, (4.15), following the general procedure described in Sect. 1.5. The first step results in the relation

$$\frac{\partial \left(c_p \rho T\right)}{\partial t} + \frac{\partial \left(\rho c_p u_k T\right)}{\partial x_k} = \frac{\mathrm{d}p}{\mathrm{d}t} + \rho \varepsilon_T - \frac{\partial h_k}{\partial x_k} \tag{4.27}$$

which, it can be observed, applies to a small open system. The final result, following details presented in Appendix A, is the equation

$$\frac{D c_v \overline{T}}{Dt} = -\overline{p}\frac{D\alpha}{Dt} + \varepsilon_M + \varepsilon \; - \frac{1}{\overline{\rho}}\frac{\partial}{\partial x_k}\left(\overline{h_k} + c_p\overline{\rho u_k' T'}\right) \tag{4.28}$$

in which D/Dt represents the change in an infinitesimal open system moving with the mean velocity. Although it simulates a parcel moving along with the mean motion, it is physically different because its mass is constantly being exchanged with the surroundings. The average rate of dissipation has been divided into two parts, ε_M the dissipation produced by the mean motion, and ε the dissipation associated with the turbulent motions. The last term represents the rate of convergence of the total heat flux. Because of the divergence theorem, this is seen to consist of the average instantaneous heat flux h_k associated with the closed system plus the turbulent flux of enthalpy associated with the turbulent exchange of air parcels across the boundary of the open system.

4.8 Energy Transformations in a Turbulent System

Following the Reynolds procedure, we average the kinetic energy per unit mass and express the result in terms of the mean and turbulent velocity components:

$$\frac{\overline{u_i^2}}{2} = \frac{\overline{u}_i^2}{2} + \frac{\overline{u_i'^2}}{2} = E_M + \text{TKE} \; . \tag{4.29}$$

The mean kinetic energy is seen to consist of the energy of the mean motion E_M and the kinetic energy associated with the turbulent motions. The latter will be called the *eddy energy* or the turbulent kinetic energy (TKE). The TKE per unit volume is seen to be minus one-half the trace of the Reynolds stress.

The average rate of change of the combined kinetic energies can be found from (4.13) by carrying through the now familiar procedure for averaging any conservation equation. The details are given in Appendix A. The result is

$$\frac{D}{Dt}(E_M + \text{TKE}) = -g\overline{w} - \frac{g}{\overline{\rho}}\overline{\rho' w'} + \overline{p}\frac{D\overline{\alpha}}{Dt} - \varepsilon - \varepsilon_M$$
$$+ \frac{1}{\overline{\rho}}\frac{\partial}{\partial x_k}(\overline{p}_{ki}\overline{u}_i + \overline{p_{ki}' u_i'} - \overline{\rho u_k' u_i'}\,\overline{u}_i - \overline{\rho u_k' u_i'^2}/2) \; . \tag{4.30}$$

Again, a small term involving pressure fluctuations has been omitted. The divergence term at the far right represents the external sources of kinetic energy, which as in the instantaneous equation (4.13), can be interpreted as the mean rate at which work is done on the boundary by the mean and eddy velocities, plus a net turbulent flux of TKE into the system. The remainder of the terms on the right side of (4.30) are internal sources of the two kinds of kinetic energy.

We obtain an equation for the rate of change of the energy of mean motion E_M by multiplying the averaged Navier–Stokes equation through term-by-term by $\overline{u}_i$ and summing over i. Again the details are found in Appendix A. The result may be written

$$\frac{DE_M}{Dt} = -g\overline{w} + \overline{p}\frac{D\overline{\alpha}}{Dt} - \varepsilon_M + \overline{u_k' u_i'}\frac{\partial \overline{u}_i}{\partial x_k} + \frac{1}{\overline{\rho}}\frac{\partial}{\partial x_k}\left[\left(\overline{p_{ki}} - \overline{\rho u_k' u_i'}\right)\overline{u}_i\right]. \tag{4.31}$$

Again, the last term on the right-hand side represents the external source of E_M. Like the instantaneous kinetic energy, this source consists of the rate at which work is done on the boundary by the total stress, which now comprises the mean molecular stress and the Reynolds stress. The remaining terms on the right side of (4.31) are the internal sources of E_M.

Finally, we can obtain an equation for the turbulent kinetic energy by subtracting (4.31) from (4.32) term-by-term. Again, the details may be found in Appendix A.

$$\frac{D(TKE)}{Dt} = -\frac{g}{\overline{\rho}}\overline{\rho' w'} - \varepsilon - \overline{u_k' u_i'}\frac{\partial \overline{u}_i}{\partial x_k} + \frac{1}{\overline{\rho}}\frac{\partial}{\partial x_k}\left(\overline{p_{ki}' u_i'} - \frac{\overline{\rho u_k' u_i'^2}}{2}\right). \tag{4.32}$$

As before, the divergence term at the far right represents the external sources of kinetic energy, and the remaining terms are internal sources of TKE.

To facilitate the discussion of these results, we can rewrite these equations in a symbolic fashion:

$$\frac{DE_M}{Dt} = B_1 + W_1 - \varepsilon_M - M + X_M \tag{4.33}$$

$$\frac{D(TKE)}{Dt} = B - \varepsilon + M + X_E \tag{4.34}$$

$$\frac{Dc_v T}{Dt} = -W_1 + \varepsilon_M + \varepsilon + X_T \tag{4.35}$$

$$\frac{D\Phi}{Dt} = -B_1 - B \tag{4.36}$$

in which the new symbols have the following definitions:

X_M X_E X_T	External source rates. These may occur with either sign and they may affect the total energy of the system.
$B_1 \equiv -g\overline{w}$	Rate of buoyant production of E_M. Reversible transformation from and to potential energy. May occur with either sign.
$B \equiv \dfrac{g}{\overline{\rho}}\overline{\rho' w'}$	Rate of buoyant production of TKE. Reversible transformation from and to potential energy. May occur with either sign.
$M \equiv -\overline{u_k' u_i'}\dfrac{\partial \overline{u}_i}{\partial x_k}$ $= \dfrac{T_{ki}}{\overline{\rho}}\dfrac{\partial \overline{u}_i}{\partial x_k}$	Rate of mechanical production of TKE; transformation from E_M into TKE but does not change the total kinetic energy. This quantity is usually positive.

$W_1 \equiv \overline{p} \dfrac{D\overline{\alpha}}{Dt}$ Reversible rate of transformation of internal energy into energy of mean motion. Does not affect the total energy. It may occur with either sign.

ε_M Always positive. Irreversibly transforms E_M into internal energy. It is usually negligible.

ε Always positive. Irreversibly transforms TKE into internal energy. Usually referred to as the dissipation rate.

Each equation has two kinds of terms, the external sources which add to (or subtract from) the total energy of the system. The remaining terms are seen to occur in each of two equations with opposite signs. Thus their function is to transform one type of energy into another type without changing the total energy of the system.

It is interesting to compare the mechanical production rate M with the dissipation term ε_M in the original kinetic energy equation (4.13). This is defined by the expression

$$\varepsilon_M \equiv \frac{\overline{\tau}_{ki}}{\overline{\rho}} \frac{\partial \overline{u}_i}{\partial x_k} \tag{4.37}$$

where τ_{ki} here refers to the viscous stress. Since M is usually positive in sign, it is tempting to think of it as a kind of dissipation. More correctly, it should be interpreted as the rate at which the energy of the organized mean motion decays into the energy of less well organized motions. Some of the energy of mean motion is transformed into internal energy directly; the rate is given by ε_M. This rate, however is vastly smaller than M. Most of the loss of E_M occurs through a transformation into TKE by way of the term M, and from there it is eventually transformed into internal energy by the turbulent dissipation rate ε. Within a homogeneous adiabatic surface layer M and ε are about equal, a fact that makes it possible to estimate ε from the more easily measured term M.

4.9 Problems

1. Devise a profile of one-dimensional motion in a fluid that yields a viscous frictional force distribution that would locally increase the kinetic energy. If no external forces are acting on the boundaries, what would be the final velocity profile? Show that total kinetic energy of the final profile is less than the total kinetic energy initially. What happened to the remainder?

2. On page 44, heat was defined so as to include only conduction and radiation. Why do we not also include convection in h_k?

3. Equation (4.18) involves an assumption that the Fourier heat conduction coefficient is constant. In soils it sometimes happens that both this coefficient and the

specific heat are functions of depth. Rewrite (4.18) so that it would apply in such circumstances.

4. Verify that (4.24) is correct under the stipulated conditions.

5. Consider a geometrically defined system that contains a star and nothing else. The star has a constant surface temperature T and radiates as a black body. Find an expression for the rate of internal production of entropy of the system.

6. Calculate the mechanical production rate of TKE at a height of 1 m for the profile II in Fig. 3.2. Estimate the value of ε at the same height. Using a value of $\nu = 1.6 \times 10^{-5} \, \mathrm{m^2 \, s^{-1}}$, calculate ε_M for the same height.

5 Diabatic Surface Boundary Layers

The distribution of the mean wind and mean concentration of passive properties in neutral surface layers has already been considered in Chap. 3. With the use of the energy equations of the last chapter, it is now possible to extend this discussion to layers that are being heated or cooled from below. We shall also consider the vertical distribution of mean temperature, which in neutral layers is constrained to have an adiabatic lapse rate.

We again confine our attention to horizontally homogeneous conditions and we assume that time rates of change are negligibly small. As before, these assumptions require that the fluxes are independent of height to a high degree of approximation. We shall also assume that the wind direction is independent of height, and define the x-axis to be in the direction of the mean motion.

5.1 Heat Flux in the Surface Layer

Rather surprisingly, even under homogeneous conditions, the horizontal component of the turbulent heat flux is not usually zero in the surface layer. However, we do not need to be concerned with its value, since it is the same everywhere; only the divergence of the flux is capable of changing the temperature. Therefore, we need only confine our attention to the vertical component, which we shall denote by the symbol H.

$$H = c_p \rho \overline{w'T'} . \tag{5.1}$$

Exchange theory cannot be applied directly because the temperature is not conservative for vertical motions. It can, however, be applied to a new variable $T_a = T + \Gamma z$, where Γz is the dry-adiabatic lapse rate equal to g/c_p or about $1°C$ per $100\,m$. T_a is conservative for adiabatic vertical motions when the air is unsaturated. At a fixed height, T_a' and T' are identical. K-theory considerations then lead to the result

$$H = -c_p \rho K_h (\partial \overline{T}/\partial z + \Gamma) = -c_p \rho \frac{\overline{T}}{\theta} K_h (\partial \overline{\theta}/\partial z) \tag{5.2}$$

in which K_h is the eddy diffusivity for heat with dimensions of length times velocity. The second form of the equation is expressed in terms of the mean potential temperature; the factor $\overline{T}/\theta$ can usually be set equal to unity. Also, within the surface layer the magnitude of the lapse rate is often so large that Γ can be ignored in comparison.

We define the following terms that will be frequently used:

Lapse	Neutral	Inversion
$H > 0$	$H = 0$	$H < 0$
Unstable lapse rate	Neutral lapse rate	Stable lapse rate
daytime	transitional	nighttime
	windy, cloudy	

5.2 The Richardson Number and the Criterion of Turbulence

We now turn to a consideration of the turbulent kinetic energy (TKE) equation as it applies in the homogeneous surface layer. This equation is

$$\frac{D\,(\text{TKE})}{Dt} = B + M - \varepsilon + X_e \tag{5.3}$$

in which B is the buoyant production rate, M is the mechanical production rate, ε is the rate of dissipation, and X_e is the rate at which TKE is increased by external sources. We shall assume for now that the last term can be neglected.

By definition, the mechanical production rate is given by

$$M = \frac{\tau}{\rho} \partial \overline{u}/\partial z \; . \tag{5.4}$$

In this and succeeding equations we use the symbol τ for the total stress, primarily the Reynolds stress, in the x-direction. By the use of K-theory, this may be written

$$M = K_m \left(\partial \overline{u}/\partial z\right)^2 \approx u_*^3/kz \; . \tag{5.5}$$

The second form of the equation is exact for neutral layers, but under diabetic conditions, the wind shear is altered as one moves away from the surface, and the expression is only approximate. The important fact to note is that the mechanical production rate decreases rapidly with height away from the surface.

The buoyant production rate is given by

$$B = -\frac{g}{\overline{\rho}}\overline{\rho'w'} = \frac{g}{\overline{T}}\overline{w'T'} = \frac{gH}{c_p\overline{\rho}\overline{T}} \; , \tag{5.6}$$

assuming that the atmosphere is dry so that the density is not affected by the humidity. In this equation, use has been made of the fact that pressure fluctuations can be neglected. Under these conditions the buoyant production is seen to be proportional to the turbulent heat flux. Its sign is therefore positive under lapse conditions, negative under inversion conditions, and zero when the layer is neutral. It is also evident that the buoyant production is nearly independent of height (the mean temperature and density change by only a minute fraction of their values in the lower 10 meters). The different rates of change with height indicate that at low levels, mechanical production tends to predominate, while at higher levels the buoyant production will have the greater influence. Another way of interpreting this result is that close to the ground we should expect profiles to be approximately logarithmic, since we have found this kind of profile to be characteristic of conditions when mechanical production plays a dominant role.

Under inversion conditions the buoyant production of TKE is negative. Turbulence must do work against the forces of gravity, a process that consumes kinetic energy. The only source of energy under these conditions is the mechanical production, and if this is insufficient to replace the losses, turbulence dies out. Richardson (1920) was the first to study this question. He reasoned that if turbulence is on the edge between occurring and or not occurring, the rate of dissipation can be neglected. Then the occurrence or nonoccurrence of turbulence would depend on the dimensionless ratio of the negative buoyancy production rate to that of the mechanical production rate. This ratio is generally called the flux Richardson number R_f. Its value is

$$R_f = -\frac{B}{M} = -\frac{gH}{c_p \overline{\rho}\, \overline{T} \tau \partial \overline{u}/\partial z} = \frac{g}{\overline{\theta}} \frac{K_h}{K_m} \frac{\partial \overline{\theta}/\partial z}{(\partial \overline{u}/\partial z)^2} \ . \tag{5.7}$$

The last form of R_f in (5.7) is obviously intended to express its value in terms of easily observable quantities but unfortunately, doing so involves us in the ratio of the exchange coefficients K_h/K_m. There is reasonably good evidence that under inversion conditions the ratio is approximately unity. One can avoid the question, as Richardson did, by defining a slightly different dimensionless number, the so called gradient Richardson number, or more simply, just the Richardson number, Ri by the equation

$$Ri = \frac{g}{\overline{\theta}} \frac{\partial \overline{\theta}/\partial z}{(\partial \overline{u}/\partial z)^2} \ . \tag{5.8}$$

Richardson argued that when this number is greater than unity, turbulence does not usually occur. More recently, observations and theory have indicated that the critical value of this ratio is smaller, probably about 0.25. This value is usually referred to as the critical Richardson number R_c. When $Ri > R_c$, observations show that turbulence is usually completely suppressed. With smaller values, including negative values, turbulence is usually present.

The absolute value of the Richardson number always increases with height. As a consequence, one normally expects that if one is close enough to the surface one can always find turbulence. The functional distribution of Richardson number with height cannot be determined without knowing the wind distribution. Therefore, it is not a useful parameter for understanding wind and temperature profiles.

It is instructive to use the approximate expression for mechanical production rate to estimate the height at which the two production rates are equal. Equating the two expressions gives this height estimate as

$$h_L = \frac{c_p \overline{\rho T} u_*^3}{kgH} \, . \tag{5.9}$$

The actual height is somewhat lower than this because (5.5) tends to overestimate the mechanical production rate at higher levels when they are influenced by buoyancy. The height defined in (5.9) is most useful in providing a length scale, the so-called Monin length

$$L = -\frac{c_p \overline{\rho T} u_*^3}{kgH} \tag{5.10}$$

which is seen to be independent of height. It should be noted that the Monin length is defined so that it has the same sign as the Richardson number, i.e. negative for lapse conditions and positive for inversions. Under neutral conditions, L is infinite.

5.3 Wind Profile Similarity

The height independence makes the Monin length an ideal length scale for expressing the height. The resulting dimensionless height is

$$\zeta \equiv z/L \, . \tag{5.11}$$

When the magnitude of ζ is small, mechanical production is large compared to buoyant production, and we should expect profiles to correspond approximately to those we have already seen under neutral conditions. In particular, under neutral conditions, the magnitude of L is infinite, and mechanical production prevails at all heights in the surface layer. The degree to which profiles depart from such neutral forms should depend on the magnitude of the dimensionless height.

This idea was first clearly stated by Monin and Obukhov in a classic paper (1954), and is now generally referred to as Monin–Obukhov similarity. This simple but powerful principle can be stated as follows: When the wind shear is properly scaled in terms of u_* and nondimensionalized, it becomes a universal function of ζ.

In simplified terms, if properly scaled, all mean wind-shear profiles are the same. Specifically, Monin–Obukhov similarity exists if

$$\frac{kz}{u_*} \frac{\partial \overline{u}}{\partial z} = \varphi_m(\zeta) \tag{5.12}$$

where φ_m is a function of ζ only. Starting with the knowledge that under neutral conditions ($\zeta = 0$), φ_m is unity, Monin and Obukhov obtained a first look at the universal function by expanding φ_m in powers of ζ and fitting the result to observations. A better fit of observational data under lapse conditions has been found to be the KEYPS equation named from the initials of five persons who independently proposed it: Kazanski, Ellison, Yamamoto, Panofsky, and Sellers. This is an implicit equation for φ_m as a function of z.

$$\varphi_m^4 - \gamma\zeta\varphi_m^3 = 1 . \tag{5.13}$$

The constant γ has a value of about 16.

A general solution for quartic equations does exist, but it is not convenient to use. Businger et al. (1971) and Dyer (1974) have suggested other equations on the basis of field experiments in Kansas and Australia. The equations most often used (usually known as the *Businger–Dyer* relations) approximate the KEYPS equation acceptably for values of z/L ranging from 0 to about -2.

$$\varphi_m = \left(1 - 16\frac{z}{L}\right)^{-\frac{1}{4}} , \quad L < 0 \tag{5.14}$$

$$\varphi_m = 1 + 5\frac{z}{L} , \quad L \geq 0 . \tag{5.15}$$

To get the mean wind as function of height, it is necessary to integrate the wind shear implied by these equations with respect to height over the range from z_0/L to z/L. Panofsky (1963) has done this as follows:

$$\overline{u} = \frac{u_*}{k} \int_{z_0/L}^{z/L} \frac{1}{\zeta}\varphi_m d\zeta \equiv \frac{u_*}{k} \left[\int_{z_0/L}^{z/L} \frac{1}{\zeta}d\zeta - \int_{z_0/L}^{z/L} \frac{1}{\zeta}(1 - \varphi_m)\,d\zeta \right] \tag{5.16}$$

or

$$\overline{u} = \frac{u_*}{k} \left[\ln\frac{z}{z_0} - \psi_m\left(\frac{z}{L}\right) \right] , \tag{5.17}$$

in which

$$\psi_m \equiv \int_0^{z/L} \frac{1}{\zeta}[1 - \varphi_m(\zeta)]\,d\zeta . \tag{5.18}$$

Analytical forms for this integral have been provided by Paulson (1970) for both the KEYPS and the Businger–Dyer functional forms. A graph of the function $\psi(z/L)$ for lapse conditions based on the Businger–Dyer formula is shown in Fig. 5.1. This graph cannot be used under stable conditions, but in such cases it is easy to use analytical means. From (5.15) it is easily shown that

$$\psi_m = -5z/L , \quad L > 0 . \tag{5.19}$$

This formula can also be used for very small values of $-z/L$ in lapse conditions since there is no discontinuity of slope at the origin..

The nature of the profiles produced by these functions can be seen in Fig. 5.2. As indicated previously, at low levels the profiles are logarithmic and approach a mean speed of 0 at a height of z_0. At some height, depending on the Monin length

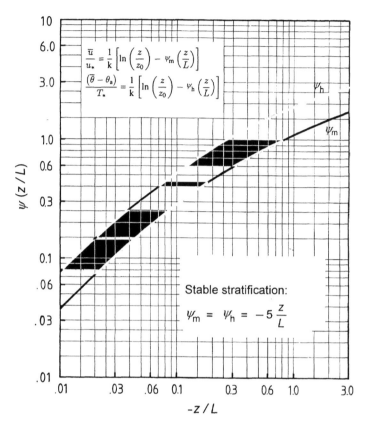

Fig. 5.1 The functions $\psi(z/L)$ evaluated from the Businger–Dyer formula for unstable stratification. See insert for stable stratification

scale, the profiles begin to deviate more and more from the logarithmic values. During lapse conditions, the wind speed and the wind shear become smaller than the neutral values, while the opposite is true under inversion conditions. Note that in this figure it is assumed that all three of the profiles shown have the same value of u_*. One should keep in mind, however, that in the daytime when lapse conditions usually prevail, the values of u_* are generally greater than at night. The typical situation is shown well in the three profiles plotted from observations in Fig. 3.2.

In going through the procedure for the determination of z_0 and u_* from observed wind profiles, it is important that the lowest two or three observed levels be situated at sufficiently small magnitudes of z/L to ensure that they lie within the logarithmic portion of the profile. Otherwise, the derived value of z_0 will be adversely affected by L. This fact can be easily demonstrated by repeating the analysis of Fig. 3.2 using the top two or three observed winds of each profile. The safest procedure is to determine z_0 for the location once and for all, using sets of

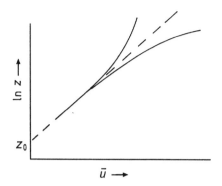

Fig. 5.2 Wind speed profiles for lapse (*left*), neutral (*dashed*), and inversion (*right*) conditions. Note that all three profiles have identical values of u_* and z_0 and that u_* is determined by the slope of the logarithmic asymptote of all three profiles

observations made under neutral conditions. Failure to recognize the importance of this caveat has caused several early analysts to conclude erroneously that the roughness length is dependent on the stability.

5.4 Profiles of Mean Temperature

Similarity theory can be applied to the non-dimensional temperature gradient in exactly the same way as has previously been done with the dimensionless wind shear. In place of the velocity scale u_* defined in (3.3), we introduce a temperature scale T_* defined in an analogous way

$$H = -c_\mathrm{p}\bar{\rho}u_* T_* . \tag{5.20}$$

The minus sign makes the sign of T_* identical to that of L and Ri. We can then proceed to define a nondimensional temperature gradient by the equation

$$\frac{\mathrm{k}z}{T_*}\frac{\partial\bar{\theta}}{\partial z} = \varphi_\mathrm{h}\left(\frac{z}{L}\right) . \tag{5.21}$$

Monin–Obukhov similarity then requires that the nondimensional temperature gradient is a function only of z/L, and once found, must be universally applicable. This form can only be determined from the analysis of field observations. For lapse conditions, these observational data are well represented by Dyer's (1967) formula

$$\varphi_\mathrm{h} = \left(1 - 16\frac{z}{L}\right)^{-\frac{1}{2}} , \tag{5.22}$$

while under inversion conditions it appears to be identical to φ_m, the form of which is (5.15).

$$\varphi_\mathrm{h} = 1 + 5z/L . \tag{5.23}$$

The mean temperature profile can be found by integrating the dimensionless temperature gradient between two levels. We usually choose the lower level z_a to be that at which the extrapolated profile is equal to the surface value of the temperature θ_a. The resulting mean temperature distribution can then be put in the form

$$\bar{\theta} - \theta_a = \frac{T_*}{k} \left[\ln \frac{z}{z_a} - \psi_h \left(\frac{z}{L} \right) \right] ,$$

(5.24)

in which

$$\psi_h \left(\frac{z}{L} \right) = \int_0^{z/L} \frac{1}{\zeta} (1 - \varphi_h(\zeta)) \, d\zeta .$$

(5.25)

The term $\psi_h(z_a/L)$ has been neglected on the assumption that z_a is small. A graph of ψ_h based on the Businger–Dyer formula is contained in Fig. 5.1. It is commonly assumed that z_a is equal to z_0, mostly for lack of a better assumption. Since the mechanisms of transfer of heat and momentum at the surface are quite different, it is likely that these two lengths are also different.

The temperature gradients close to the surface may be very large if the surface heating rate is large. This gradient can be calculated by solving (5.21) for $\partial \bar{\theta}/\partial z$. Over a strongly heated surface, H can easily be 600 W m^{-2}, and with a light wind, u_* might be 0.4 m s^{-1}. Using typical values of the other quantities one finds $L = -10$ m and $T_* = -1.25°$C. The gradient at a height of 10 cm is then found to be $-34.0°$C m^{-1} or about 3400 times the dry adiabatic rate. Under these conditions there is a strong upward increase of density with height, and light entering this layer nearly horizontally is strongly refracted upward, creating the possibility of mirages.

The vertical distribution of other quantities near the ground can be obtained in a similar way. For example with water vapor one can define a vapor concentration scale q_* using the evaporation rate E in place of H:

$$E = -\bar{\rho} u_* q_*$$

(5.26)

and set the nondimensional gradient of $\bar{q}$ equal to a function $\varphi_q(z/L)$;

$$\varphi_q \left(\frac{z}{L} \right) \equiv \frac{kz}{q_*} \frac{\partial \bar{q}}{\partial z} .$$

(5.27)

One then obtains an equation for the profile of q,

$$\bar{q} - q_a = \frac{q_*}{k} \left[\ln \frac{z}{z_a} - \psi_q \left(\frac{z}{L} \right) \right] .$$

(5.28)

A practical use of this equation is to solve it for the evaporation rate E:

$$E = -C_q \bar{\rho} \bar{u} (\bar{q} - q_a)$$

(5.29)

in which the bulk coefficient C_q is equal to

$$C_q = \frac{k^2}{(\ln(z/z_a) - \psi_q)(\ln(z/z_0) - \psi_m)} . \tag{5.30}$$

The value of q_a is usually taken to be the saturation specific humidity at the temperature of the water surface, while $\bar{q}$ is put equal to the mean prevailing specific humidity at height above the surface. The value of C_q cannot be calculated easily because the value of z_a is unknown. We therefore have to rely on measurements of the coefficient for actual values of E. However, (5.30) is useful in telling us something about the effects of wind speed and heating rates on the rate of evaporation. For ψ_q we can usually use the value of ψ_h. For a specific height both ψ_q and ψ_m are functions of the heating rate. Under unstable conditions, both values of ψ are positive and the coefficient C_q is greater than its neutral value. Conversely, under stable conditions the evaporation rate is smaller than in neutral conditions for the same wind speed and water vapor concentrations.

We may look again at the power law expression for the wind profile discussed in connection with Fig. 3.3. Because of the reciprocal relation between K_m and the wind gradient that is implied by the constancy of the stress, the effect of lapse conditions is to reduce the wind shear at upper levels. Since the wind is normalized to the value u_1 at a height of z_1, reducing the shear at higher levels must be accompanied by an increase of shear in the lowest levels. The result is a reduction of the exponent p for lapse conditions and conversely, an increased value with inversions. Brunt (1952, p. 252) has reported values as low as 0.01 for lapse conditions and as high as 0.62 for inversions. It can easily be shown that the value of p is given by

$$p = \frac{\varphi_m(z/L)}{\ln(z/z_0) - \psi_m(z/L)} . \tag{5.31}$$

The dependence of this expression on height shows that the power law is not perfectly compatible with Monin–Obukhov similarity.

5.5 Some Useful Relationships

From the definitions of and we have the useful relations

$$K_m = \frac{ku_* z}{\varphi_m} \quad \text{and} \quad K_h = \frac{ku_* z}{\varphi_m} \tag{5.32}$$

and the resulting relation

$$\frac{K_h}{K_m} = \frac{\varphi_m}{\varphi_h} . \tag{5.33}$$

Use of the Businger–Dyer formulas now allows us to investigate the relation between K_h and K_m. The result is

$$\frac{K_h}{K_m} = \left(1 - 16\frac{z}{L}\right)^{\frac{1}{4}} \tag{5.34}$$

for lapse conditions. Since z/L is less than 0 under these conditions, the ratio K_h/K_m becomes increasingly greater than unity under progressively greater values of $z/-L$. When $z/-L$ is 1, K_h is about double the value of K_m. Under inversion conditions, φ_m and φ_h are equal, and we can treat the two Richardson numbers R_f and Ri as equal in this situation. Information about exchange coefficients for other properties is quite sketchy. What there is indicates that K for passive properties is about equal to K_h. The same can also be said for the values of φ and ψ. Thus the evidence suggests that momentum is different from other properties, presumably because of associated pressure fluctuations that prevent it from conserving itself during the mixing process.

We have assumed up to this point that density fluctuations depend only on temperature fluctuations. When moisture fluctuations are also present, the density fluctuations depend on the fluctuations of virtual temperature instead of the actual temperature, as was pointed out in (4.26). As a result, the buoyant production of TKE is also a function of the mean evaporation rate according to the equation

$$B = \frac{g}{\overline{\rho T}}\left(\frac{H}{c_p} + 0.61\overline{T}E\right) \tag{5.35}$$

in which E is the evaporation rate in $\mathrm{kg\,m^{-2}\,s^{-1}}$.

Finally we can mention the relation between R_f and z/L:

$$z/L = \varphi_m R_f . \tag{5.36}$$

Under stable conditions ($0 < Ri < R_c$), this leads to the relation

$$Ri = \frac{z/L}{(1 + 5z/L)} . \tag{5.37}$$

5.6 Problems

1. Show that the mechanical production rate is given exactly by

$$M = u_*^3 \varphi_m/kz . \tag{5.38}$$

If L is $-10\,\mathrm{m}$, at what height are the mechanical and buoyant rates actually equal? (Note: An iterative procedure will be needed to arrive at the value.)

2. Find an expression for the ratio of the prevailing wind shear and that which would prevail with the same under neutral conditions. Relate this ratio to the dimensionless function φ_m.

3. In the light of the previous question, define φ_m and φ_h in words.

4. Explain without using equations why: (1) the wind shear must decrease with height; (2) why the wind shear is less at most levels under unstable conditions

than under neutral conditions for the same value of u_*; and (3) why the difference described in (2) increases with height.

5. Revisit the TKE equation assuming that the Richardson number approaches a limit of 0.25 as the TKE dies out, rather than unity as originally assumed by Richardson. Assuming that external sources of energy can be ruled out, what does the existence of a critical Richardson number imply about the dissipation rate?

6. When the negative buoyant production during inversion conditions becomes large compared to the mechanical production, we expect turbulence to die out. What is the limiting value of z/L under these conditions? What value of R_c is implied by the Businger–Dyer formulas?

6 Homogeneous Stationary Planetary Layers

We have previously defined the planetary boundary layer (PBL), also known as the atmospheric boundary layer (ABL), as the entire layer of the atmosphere that has its properties directly influenced by contact with the earth's underlying surface. Though some turbulence occurs from time to time in the free atmosphere, the fluxes that we have been concerned with are primarily due to interactions with the earth's surface, and their values decrease in magnitude to near zero at the top of the PBL. While we have found it possible to ignore the vertical variation of flux with height in the surface layer, however, we cannot do so in the more extended upper or outer portion of the PBL.

In the initial approach to the study of this extended layer, we shall continue to make the simplifying assumptions of horizontal homogeneity and stationarity. Over continents it is difficult to justify the assumption of stationarity because the large amplitude of the diurnal changes implies variations of the fluxes with height that rival those of a stationary PBL. This simplification, however, provides a useful introduction to some of the important properties of the layer, and the effects of time rates of change will be fully considered in the next chapter.

The existence of true stationarity implies the lack of most fluxes at the surface. This implication results from the fact that nearly all of the properties we are concerned with have no sources or sinks within the boundary layer. Since we assume that there is no flux through the upper boundary, stationarity demands that there be no flux through the lower boundary either, and in addition no change of flux with height. For example, if there were a significant heat flux at the surface, the temperature of the layer would increase with time; thus stationarity implies a neutral stratification. Momentum is an exception because the Coriolis force is a momentum source, and stationarity implies a change of momentum flux with height. Typically, the flux is zero at the top of the Ekman layer and has its largest magnitude at the surface. Although the requirements of true stationarity are seldom completely met, the deductions we make are a useful approximation to typical conditions.

The equations of motion for horizontally homogeneous mean conditions can be written

$$\frac{\partial \overline{u}}{\partial t} = f\overline{v} - \frac{1}{\overline{\rho}}\frac{\partial \overline{p}}{\partial x} + \frac{1}{\overline{\rho}}\frac{\partial \tau_x}{\partial z} \tag{6.1a}$$

$$\frac{\partial \overline{v}}{\partial t} = -f\overline{u} - \frac{1}{\overline{\rho}}\frac{\partial \overline{p}}{\partial y} + \frac{1}{\overline{\rho}}\frac{\partial \tau_y}{\partial z} \tag{6.1b}$$

in which f is the Coriolis parameter,

$$f \equiv 2\Omega \sin \phi . \tag{6.2}$$

Ω is the angular velocity of the Earth's rotation, and τ_x and τ_y are the components of the tangential stress exerted across a horizontal surface. It is seen that because of the pressure gradient and Coriolis forces, a steady state is possible even though a variable field of stress is present.

The geostrophic wind is defined as that wind that is reached in a steady state when there is no friction. Thus, if u_g and v_g represent the geostrophic components we have

$$fv_g \equiv \frac{1}{\overline{\rho}}\frac{\partial \overline{p}}{\partial x} \tag{6.3a}$$

$$fu_g \equiv -\frac{1}{\overline{\rho}}\frac{\partial \overline{p}}{\partial y} , \tag{6.3b}$$

and we can use these free-stream wind components to replace the pressure gradient terms in the equations. Thus the equations for stationary motion become

$$\frac{1}{\overline{\rho}}\frac{d\tau_x}{dz} = -f\left(\overline{v} - v_g\right) , \tag{6.4a}$$

$$\frac{1}{\overline{\rho}}\frac{d\tau_y}{dz} = f\left(\overline{u} - u_g\right) . \tag{6.4b}$$

The departure of the wind from geostrophic, which appears on the right side of these equations, is often called the *velocity defect*. The equations suggest that if the wind and the horizontal pressure gradient can be observed as functions of height, it should be possible to derive estimates of the vertical distribution of stress above the surface layer. This method is known as the *geostrophic departure* method, and has been used with apparent success by Lettau (1950, 1957). As we have seen in Chap. 3, good estimates of the stress direction and magnitude at the top of the surface layer can be made from measured surface layer wind profiles, although it is not clear that such values are representative of any large area. The geostrophic departure method consists of integrating the wind departures with respect to height from the surface layer to other heights within the planetary boundary layer. Unfortunately, there are many practical difficulties in applying this method. The local value of the pressure gradient cannot usually be determined with sufficient accuracy: the measured winds are seldom accurately representative of the larger-scale mean wind; the assumption of stationarity is not accurately fulfilled; and finally, the geostrophic departures themselves tend to be small above a few hundred meters.

Because of the difficulties of measuring the stress in the layers above the surface layer, we know very little from direct or indirect observation about the distribution of stress throughout the planetary boundary layer, or about derived quantities such as the kinematic eddy viscosity K_m. Ellison (1955) has solved the equations assuming that K_m continues to increase upward linearly as it does in the surface layer. The surface drag predicted by Ellison's solution is too small. The linear increase of K_m in neutral surface layers gives way to a more nearly constant value at greater heights and there is some opinion that it decreases to zero near the top of the boundary layer. Lettau's analysis of Mildner's wind observations at Leipzig indicated a maximum value of $15\,m^2\,s^{-1}$ at about $250\,m$. Before one gets to the levels where a definitive clarification of this uncertainty can be given, the geostrophic departure becomes too small to give a trustworthy result.

6.1 The Ekman Spiral

The solution we shall discuss in this section was originally derived by Ekman (1905). Ekman was an oceanographer, and his solution was for ocean drift currents – currents imposed by the stresses exerted on the ocean by the atmosphere. Except for the difference of density between the atmosphere and the ocean, and the preference of some oceanographers to define z so that it increases downward, the dynamics of the ocean are quite similar to atmospheric dynamics. The solution was derived independently for the atmosphere by Taylor (1915) using boundary conditions that are more applicable to this medium.

We shall follow closely the development of Taylor, except that we shall employ the use of scaled, dimensionless variables. By doing so, we reduce the number of independent variables to a minimum. What is even more important, we make it possible to employ intuitive reasoning in the induction of a key relationship that is needed to make the result universal.

It can be seen from (6.4) that if there were no drag of the wind at the surface, the wind would be geostrophic at every level. Thus we should expect the velocity defects to scale as u_*; that means that when expressed in units of u_*, the magnitude of the terms of (6.4) is of order unity in the portion of PBL where they are large enough to be significant.

Without much justification, Ekman and Taylor made the assumption that K_m is independent of z. After many attempts to improve on Ekman, most scientists have concluded that this assumption is about as good as any that can be made at the present time. We shall also assume that the geostrophic wind is either constant or at most a linear function of height, so that its second derivative with respect to height is zero. Accordingly we have

$$\frac{d^2}{dz^2}\left(\overline{u} - u_g\right) = -\frac{f}{K_m}\left(\overline{v} - v_g\right) \tag{6.5a}$$

$$\frac{d^2}{dz^2}\left(\overline{v} - v_g\right) = \frac{f}{K_m}\left(\overline{u} - u_g\right) \ . \tag{6.5b}$$

For the scaled velocity defects, we let

$$U \equiv \frac{\overline{u} - u_g}{u_*} \tag{6.6a}$$

$$V \equiv \frac{\overline{v} - v_g}{u_*} . \tag{6.6b}$$

For height, we use units of u_*/f to get a dimensionless height

$$Z \equiv fz/u_* \tag{6.7}$$

and we introduce a dimensionless parameter β defined by

$$\beta \equiv \sqrt{\frac{u_*^2}{2fK_m}} . \tag{6.8}$$

The equations for the velocity defects can then be written

$$\frac{d^2U}{dZ^2} = -2\beta^2 V \tag{6.9a}$$

$$\frac{d^2V}{dZ^2} = 2\beta^2 U . \tag{6.9b}$$

What we have done is to reduce the number of variables to a set of pure numbers of order unity which satisfy a pair of second order ordinary differential equations. The solution of these equations will enable us to plot a universal vector diagram for the velocity defects that can be used for all geostrophic winds, all latitudes, all values of u_*, and all values of K_m. We shall see, however, that not all of these variables can be selected independently.

The solution of these equations and the geometrical interpretation of the solution can both be greatly simplified by introducing the complex variable W defined by

$$W \equiv U + iV \tag{6.10}$$

where $i^2 = -1$. The two equations (6.9) then become a single equation

$$\frac{d^2W}{dZ^2} = 2i\beta^2 W . \tag{6.11}$$

The solution of this equation is

$$W = W_0 \exp\left[-(1+i)\beta Z\right] \tag{6.12}$$

as can easily be verified by substituting it into the differential equation (6.11). The number W_0 is a constant of integration, and it can be seen to be the value that W has when Z is 0. There is also a second solution of this equation in which the minus sign in the exponent is replaced by a plus sign. This solution is unacceptable on physical grounds, because it would lead to an infinitely large velocity defect if Z increases without limit.

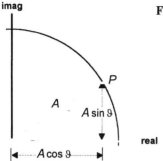

Fig. 6.1 Graphical representation of the number P

Let us review the geometrical interpretation of a complex number such as $P = A \exp(i\vartheta)$ where A and ϑ are real numbers. We remember the identity $\exp(i\vartheta) \equiv \cos\vartheta + i\sin\vartheta$. Therefore, on the complex plane shown in the figure, we see that the point P is situated on a circle of radius A at a point that makes an angle ϑ in a counterclockwise direction from the real axis.

Likewise, the number $R = Pe^{i\varphi} = Ae^{i\vartheta}e^{i\varphi} = Ae^{i(\vartheta+\varphi)}$ is seen to represent a point on the same circle but rotated at an angle φ to the left of the direction of P. Thus multiplication by $e^{i\varphi}$ simply changes the direction of the radius vector on the complex plane without changing the magnitude. If φ is positive, the rotation is toward the left, while if negative, it is to the right.

We can now return to the interpretation of the solution (6.12), which is identical to

$$W = \left(W_0 e^{-\beta Z}\right) e^{-i\beta Z} . \tag{6.13}$$

The vector W_0 is the value of W at $Z = 0$. Its magnitude and direction depend on the boundary condition which will be discussed later; for now we shall assume that its direction is rotated 135° to the left of the real axis. We can leave its magnitude undefined for the present. The magnitude or length of W at any height Z greater than zero can be found by multiplying the magnitude of W_0 by $e^{-\beta Z}$, a number that is less than unity and gets smaller and smaller the greater Z becomes. The limiting value as Z approaches infinity is zero. The direction of W is determined by $e^{-i\beta Z}$, which represents a rotation to the right through an angle βZ. Combining these results for values of Z increasing indefinitely from zero to infinity leads to the spiral shown in Fig. 6.2.

The curve shown in Fig. 6.2 is often called the *Ekman spiral*; generically, it is an equiangular spiral curve whose tangent at any point makes a 45° angle with the intersecting radius vector. When βZ is equal to the number π, the direction of W is opposite to its surface value, and its magnitude drops to about four percent of the surface magnitude. Each further increase of βZ by π results in a further rotation of 180° and a reduction of magnitude by another factor of about 25. Thus it is seen that the velocity defect virtually disappears after doing less than one complete rotation. The rate at which it does so with increasing actual height depends on the magnitude of β and u_*.

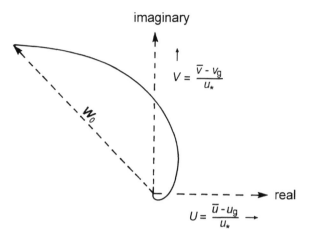

Fig. 6.2 Scaled geostrophic wind defect hodograph. W_0 is the surface value

The actual wind at any height is not indicated by the solution shown in Fig. 6.2. However, if the true wind can be specified at any one height, it is determined at all other heights. Taylor did this by assuming that $Z = 0$ corresponds to height of surface wind observations (usually about 10 m), and assuming that the direction of the wind is parallel to the tangent of the spiral curve at this level. Thus, in effect the origin of the actual wind vectors is located on a tangent extending to the left of the spiral in Fig. 6.2, a distance equal to the speed of the surface wind scaled by u_*. The actual wind at any level above 10 meters would then be found by connecting this point of origin with the point on the curve corresponding to the scaled height Z.

What is the justification for choosing to match the wind at a height of 10 meters rather than at some other level, or perhaps the true surface? The best justification seems to be that it works quite well over the terrain of southern England where the observations Taylor worked with were made. Selecting the ground surface is not acceptable because it results in a surface wind that always blows at 45° to the surface isobars and a speed at ten meters height that is far too small.

We have scaled Z so that its value is measured in units of u_*/f. Typically, u_*/f is about 10 000 meters. In these units, the roughness of the surface is many orders of magnitude smaller than 1. In fact it is $z_0 f/u^*$. The reciprocal of this number is often called the *surface Rossby number*. (Rossby actually used a similar number in a paper written jointly with R. B. Montgomery (1935), but the name has been coined by others more recently.) Typical values of this surface Rossby number range from 10^4 to 10^9, depending on the surface roughness, the latitude, and the wind speed. The number is so large that for all intents and purposes it may be treated as infinity.

The implication of this situation is that in effect, the wind distribution in the vicinity of the boundary is indeterminate if heights are measured in units of u^*/f.

If this is so, how can we hope to specify the boundary condition of the Ekman solution?

There is another problem. We have specified that in the PBL, the exchange coefficient K_m is constant. Yet, we have seen earlier that in the surface layer K_m is a linear function of height, and its value at each height is calculable from a knowledge of the surface stress. How can the solution be reconciled to a boundary condition when the conditions that were used to derive it are not satisfied in the immediate vicinity of the surface?

A general discussion of this dilemma has been given by Blackadar and Tennekes (1968). Using *matching theory* it is possible to deduce the characteristics of the flow in the outer (Ekman) layer and the inner (surface) layer without making any assumptions about the distribution of exchange coefficients. The velocity defect distribution in the outer layer is independent of the surface Rossby number when scaled as described above. This situation is called Rossby number similarity. In the inner layer, the velocity (not the defect) also scales as u_*, but the height must scale with z_0. Matching of these flows to the irreconcilable height scales results in a pair of equations that enables the wind in the surface layer to be determined from the free-stream velocity, i.e. from the isobars on a weather map. A brief review of this theory is provided in Appendix C.

6.2 A Two-Layer Model of the PBL

A different way out of the dilemma is to look at a simple two-layer model that has been discussed previously by the author (1974). With it we can provide a simple description of the relation between the flow in the surface layer and the geostrophic wind determined from the pressure distribution on a weather map. Such questions as the effect of latitude, surface roughness, and stability on the speed and direction of the surface wind relative to the isobars can readily be answered with this model. To avoid unimportant detail, we shall assume neutral conditions in the following discussion.

The lower of the two layers is based at the ground and corresponds to what we have previously referred to as the surface or the inner layer. In this layer the velocity is described by the logarithmic equation (3.6), and the eddy viscosity K_m increases linearly in the upward direction in accordance with (3.4). It is convenient to define the x-direction as the direction of the wind in this layer We denote the height of the top of this layer as h, to be designated later. At this height we have

$$u_h = \frac{u_*}{k} \ln \frac{h}{z_0} \, , \quad v_h = 0 \, , \quad K_m = ku_*h \, . \tag{6.14}$$

The upper layer of the model consists of the Ekman layer that we have previously considered. The height variable Z is now the height above the interface, not the height above the ground. Within this layer, K_m is constant, and has the value ku_*h, the value at the top of the surface layer. In this way the continuity of

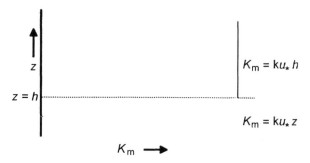

Fig. 6.3 Architecture of the two-layer model of the boundary layer

K_m across the interface is assured. It is also necessary to assure that the velocity vector and its vertical derivative are continuous across the interface, and satisfying this requirement, in effect, enables one to satisfy the boundary condition of the Ekman spiral and derive the relation between the wind flow in the two layers.

The architecture described above is pictured in the diagram in Fig. 6.3. The interface at height h is common to both layers. In the surface layer it is expressed in units of z_0, while in the outer layer it must be expressed in units of u_*/f. The ratio of these is the surface Rossby number $u_*/f z_0$. The reciprocal of the surface Rossby number is $Z_0 = f z_0/u_*$, or the value of the surface roughness parameter expressed in units of u_*/f.

To complete the joining of the two layers, it is necessary to define the height h. Rossby number similarity requires that the flow in the outer layer be dependent only on Z_0. Therefore, we cannot specify h in units of z_0. We must scale h in units of u_*/f. The simplest hypothesis we can choose is

$$h = \gamma \, u_*/f ,$$ (6.15)

where γ is some constant yet to be determined. Substitution into (6.8) and (6.14) gives

$$K_m = \gamma k u_*^2/f \quad \text{and} \quad \beta = \sqrt{\frac{1}{2k\gamma}}$$ (6.16)

Thus K_m is completely specified at every height. The parameter β in the solution for the velocity defect in the outer layer is a pure numerical constant.

The x-direction has been defined as the direction of the wind in the surface layer. This is also the direction of the stress vector everywhere in the lower layer including the upper interface surface. The stress vector in the upper layer is in the same direction as dW/dZ and therefore, parallel to the tangent of the Ekman spiral. To ensure continuity of the stress vector at the interface, this tangent must point in the direction of the real axis, since this axis is also in the x-direction. The result is the configuration we have already looked at in Fig. 6.3. This means that the direction of W_0 is required to make an angle of 135° to the real axis. The magnitude of W_0 follows by making the magnitude of the two wind shears the same at the interface. Doing so leads to the result

$$|W_0| = \sqrt{2}\beta .$$ (6.17)

Thus W_0 is completely determined and is a pure numerical constant that depends only on the undetermined constant γ. This constant can only be determined by comparison of the model predictions with observation, and these will be described later. The value that works best has been found to be 0.01. Using this value, we find

$$\beta = 11.2 \quad \text{and} \quad |W_0| \cong 16 .$$ (6.18)

The two solutions have now been completely determined. In the upper layer, we know only the scaled velocity defects $(\overline{u} - u_g)/u_*$ and $(\overline{v} - v_g)/u_*$ as functions of Z, the distance above the interface scaled by u_*/f. In the lower layer we know only the scaled velocity, not the velocity defect, as a function of height scaled by z_0. To facilitate the comparison, we rescale the heights in the lower layer by u_*/f so that all the velocities can be placed on a two-dimensional plane. This plane is pictured in Fig. 6.4.

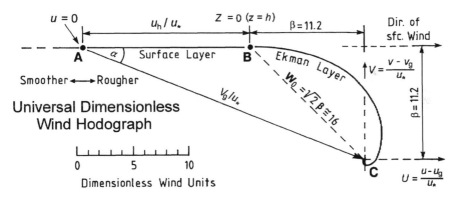

Fig. 6.4 Universal dimensionless planetary boundary layer wind hodograph. The location of the wind vector origin at the surface, point **A**, depends on the surface roughness but is always on the extended line passing from **B** through **A**

6.3 Universal Wind Hodograph and the Resistance Laws

We refer to the dimensionless velocity vector diagram shown in Fig. 6.4. Displacements on this diagram are pure numbers, and the scale is indicated in the lower left space. The key point on this diagram is the point labeled **C**. This point represents the origin of the velocity defect distribution we have been calling the Ekman spiral. A second key point is that indicated at **B**, the velocity defect at the interface. Note that the U-axis has been made parallel to the wind shear vector at **B**. As we have seen, the separation of the points **B** and **C** is then specified by the complex number W_0, which has a magnitude of about 16.

As Z increases with height above the interface, the end of the wind defect vector moves along the spiral curve from **B** toward **C**. The wind defect vector is found by rotating the direction of W, the vector connecting **C** to the point on the spiral, toward the right through the angle βZ in radian measure. The actual wind vector itself, however, cannot be determined because its origin is the wind at the ground, which has not yet been located.

We arrive at the location of the ground by proceeding downward from the interface location **B**. The surface layer wind lies along the previously defined x-direction, the direction of the line connecting **A** and **B**. The origin of the wind vectors at all heights is the point **A**, which is distant from **B** by an amount u_h/u_*. From (6.14) this distance is

$$\frac{u_h}{u_*} = \frac{1}{k} \ln \left(\frac{0.01 u_*}{f z_0} \right) = \frac{1}{k} \ln \left(\frac{0.01}{Z_0} \right) ; \tag{6.19}$$

the distance of this point from **B** depends on the roughness of the surface.

The wind vector at any height is the vector that connects the point **A** with the location on the universal hodograph, the curve connecting the points **A**, **B**, and **C**. Starting at the ground, the wind increases along the line **AB** logarithmically without changing direction. After the point **B** is reached the wind vector traverses the Ekman spiral in the fashion previously described, approaching the point **C** at infinite height. The vector **AC** is the free-stream velocity scaled by u_*. It is seen to make an angle to the surface wind direction α that depends on the universal surface roughness number Z_0, which is also the reciprocal of the surface Rossby number. With smoother surfaces in such units, the angle α is smaller than with rougher surfaces, and the value of u_* becomes a smaller fraction of the geostrophic wind.

Under neutral conditions the equations for the components of the scaled geostrophic wind vector can be written down from the dimensions pictured in the diagram,

$$\frac{u_g}{u_*} = \frac{1}{k} \left(\ln \frac{u_*}{f z_0} - A \right) \tag{6.20a}$$

$$\frac{v_g}{u_*} = -\frac{B}{k} \tag{6.20b}$$

in which A and B are constants with the values

$$A = -k\beta - \ln(0.01) \cong 0 \tag{6.21a}$$

$$B = k\beta \cong 4.5 . \tag{6.21b}$$

These equations were first derived by Kazanski and Monin (1961) and are known as the *resistance laws*. They are quite general and are not restricted to the assumptions on which the two-layer model is based. With their use, both the values of u_* and α can be determined from the surface roughness, the latitude, and the geostrophic wind. The values of A and B predicted by the model are functions of the constant γ, and the value quoted above gives the best agreement of A and B with observations.

Usually one likes to use the resistance laws to determine u_* and α when the geostrophic wind direction and speed are known. This is awkward because the resistance laws give V_g/u_* and α explicitly as functions of the unknown u_* rather than V_g. However, (6.20) implicitly relates u_* and α to V_g/fz_0, a modified form of the surface Rossby number. These relations, shown in (6.23), and in Fig. 6.5, permit the unknown quantities to be determined directly when the geostrophic wind, latitude, and surface roughness are known. These curves are universal in the sense that they involve pure numbers of scaled variables. However, it should be remembered that strictly, they apply only to neutral barotropic conditions.

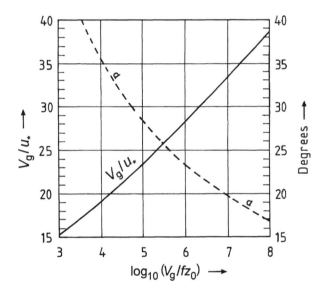

Fig. 6.5 Resistance laws for neutral conditions. $A = 0$, $B = 4.5$

We may now look at how these curves can be extended to diabatic conditions. A diabatic PBL implies temperature changes. Therefore it is theoretically incompatible with a steady state. Usually, however, the rates of change of temperature are small enough to allow us to neglect the vertical variation of the fluxes without serious error. The two-layer model can then be applied as before, but it must be modified so as to be consistent with the more general provisions of Monin–Obukhov similarity in the surface layer. Specifically, Monin–Obukhov similarity influences the values of K_m, wind speed, and stress at the top of the surface layer. The effect of these changes is to make A and B functions of a dimensionless

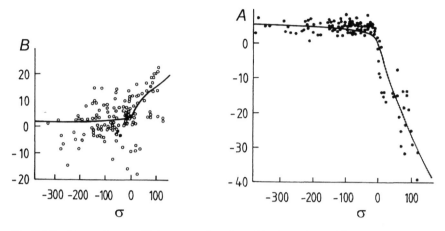

Fig. 6.6 Resistance law coefficients as a function of hydrostatic stability

stability parameter σ, defined by

$$\sigma \equiv ku_*/fL . \tag{6.22}$$

The resulting relation between A and B and this parameter, as found by the author, is shown in Fig. 6.6.

Under unstable conditions, A and B are nearly independent of surface heating. This is probably because under such conditions, heights do not scale as u_*/f but rather as the height of the mixed layer itself. Under stable conditions, both A and B are strongly dependent on the stability parameter. The effect of these variations is to make the ratio of u_*/V_g much smaller under stable conditions than in neutral conditions. This conclusion seems to be supported both by the predictions of the model and the observations of A vs. σ, to which the ratio V_g/u_* is most sensitive. The angle α is more sensitive to the value of B. Theory predicts that B increases with increasing stability, tending to cause the angle α to increase. Observations of the effect of σ on α are inconclusive. It appears that the angle between the wind direction and the geostrophic wind is extremely sensitive to local inhomogeneities and the effects of baroclinicity.

Baroclinicity causes the geostrophic wind speed and direction to change with height. The effect of this factor has been reported by Clarke and Hess (1974). One effect is quite prominent. Cold advection is usually associated with instability up to considerable heights. This combination is usually associated with very large angles of α, often approaching 90°.

6.4 The Mixed Layer of the Ocean

The stress at the surface that causes velocity defects in the atmosphere also causes velocity defects in the ocean. When the geostrophic current is zero, the velocity defect is the actual velocity and is called the drift current.

In general, the velocity defects in the ocean scale in the same way and we should expect them to satisfy the uniform dimensionless hodograph shown in Fig. 6.4. The value of u_* in the ocean is quite different from that in the atmosphere because of the density difference. Across the air-ocean interface the stress is continuous. For this reason the ratio of the value of u_* is the square root of the reciprocal ratio of the densities, or about 1 to 29. Thus the drift currents are much smaller than the winds in the atmosphere that create them.

Traditionally it has been assumed by oceanographers that K_m is independent of depth. Perhaps because of the large orbital motions of waves, the existence of a logarithmic layer close to the surface does not appear to have been observed, even though similarity with atmospheric flows suggest that it ought to exist. We must presume then that the ocean surface is located at **B** on the universal hodograph, and the origin of the drift current velocity vectors is located at **C**, the velocity at infinite depth. The shear vector of the drift current at the surface is tangential to the spiral at **B** and is directed from right toward left in the minus x-direction. This must be the direction of the stress exerted on the surface as well as the direction of the wind just above the surface. Thus it is seen that the surface drift current makes an angle of 45° to the right of the direction of the surface wind. The resulting hodograph of the drift current at various depths is shown in Fig. 6.7. Because of the Earth's rotation, the total steady-state mass transport of the drift current in a deep ocean is at right angles to the direction of the surface wind stress (see Problem 6).

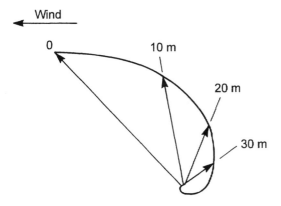

Fig. 6.7 Typical distribution of drift currents in the ocean boundary layer

The depth of the PBL in the atmosphere is usually defined as the height at which the wind direction first becomes parallel to the geostrophic wind. This parallelism usually occurs at about $0.25u_*/f$ or typically at about $1000\,m$. Because of the different value of u_* that prevails in the ocean, the typical depth of the mixed layer is around $35\,m$. If the logarithmic surface layer found in the atmosphere actually exists in the ocean its depth would probably be less than $1\,m$.

6.5 Problems

1. Show that the following equations can be used to relate u_*/V_g and α to the modified surface Rossby number V_g/fz_0.

$$\ln\left(\frac{V_g}{fz_0}\right) = A + \ln\left(\frac{V_g}{u_*}\right) + \left(\frac{k^2 V_g^2}{u_*^2} - B^2\right)^{\frac{1}{2}} \tag{6.23a}$$

$$\sin\alpha = \frac{B}{k}\frac{u_*}{V_g}. \tag{6.23b}$$

2. Anticipating the next chapter, we find that over continents, the wind often varies by $10\,m\,s^{-1}$ between day and night at some elevations. Assuming that this variation is sinusoidal, estimate the order of magnitude of $\partial\bar{u}/\partial t$ and compare it with the other terms in the equation of mean motion (6.1).

3. From the definition of the geostrophic wind, how can one justify its use in equations such as (6.4) where friction plays an important role?

4. Is it correct to say that the motion in the ocean is caused by friction? If so, how can one reconcile this statement with the second law of thermodynamics?

5. What is the magnitude of the Prandtl mixing length of the upper layer of the two-layer model?

6. Find a relation between the total transport of the drift current (mass transported across a unit width of the current from the surface down to infinite depth) and the surface stress. What is the direction of this total transport vector?

7. Show that the curl of the surface stress on the ocean is equal to the horizontal divergence of the of the total drift transport vector.

7 Unconstrained Boundary Layers

In the real world it is almost impossible to find horizontal surfaces that are truly homogeneous; nor is it usually possible over continents to find a truly steady state condition that persists long enough to satisfy the restrictions we have imposed up to now. Simplifications such as we have made serve a useful purpose in permitting us to reach an introductory understanding, but applications as well as a comprehensive understanding of the atmosphere require that we consider more complex systems.

Inhomogeneous surfaces come literally in all shapes and sizes and, when combined with the presence of time rates of change, are so complex that each realization needs to be handled by a three-dimensional time-dependent model which is smaller in scale, but comparable in complexity to the global circulation models used to predict the weather. Such models have become possible with the advent of large computers, but their use is limited by uncertainties in the closure methods, which are generally more difficult to prescribe than in the simple homogeneous environments.

7.1 Flow Downwind of a Change of Roughness

One of the simplest kinds of inhomogeneity is a sudden change of surface roughness. The wind flow in this situation has been extensively studied, both theoretically and through observation.

We may suppose that the wind has reached an equilibrium in crossing a rough surface and suddenly encounters a smooth one. We shall suppose the line of demarcation between the two surfaces is perpendicular to the wind direction so that the distribution of all properties is a function of x and z only.

With a neutral stratification, the wind profile is logarithmic over the rough surface and the stress is independent of height. As the air moves over the smooth surface, the lowest levels are affected immediately, while higher levels are only affected some time later. One can define a plane separating the two regions. The plane intersects the ground at the line of demarcation and slopes upward over the new area. Above this plane, the air has not yet been affected by the new surface. Below the plane, the flow is in a state of transition; it is accelerating and the stress varies with height. This plane is indicated schematically in Fig. 7.1 by the line AB_2. Its gradient is usually about 1/10.

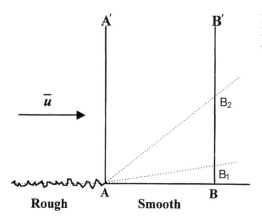

Fig. 7.1 Configuration of towers **A** and **B** in the region downwind of a sudden change of surface roughness

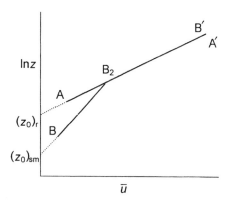

Fig. 7.2 Wind profiles **AA′** and **BB′** observed on towers **A** and **B** at, and downwind of, an abrupt change of surface roughness

If a mast is erected at point **A** of the upper figure, the observed wind profile is shown by the line **AA′** in the lower Fig. 7.2. The projection of the line to $\bar{u} = 0$ is the roughness of the original surface. Winds observed on a mast located at **B** generally form a line **BB₂B′**. The upper segment is similar to that observed at mast **A**. The lower segment is approximately logarithmic, and if extended downward intersects $\bar{u} = 0$ at a height appropriate to the new roughness. Unlike the equilibrium profiles we have looked at before, however, the gradient of this logarithmic segment cannot be used to get u_*. The reason is that the winds in the lower region are accelerating.

There is a second surface, indicated in the figure by the dashed line **AB₁**, below which the adjustment to the new surface may be considered to be complete. Below this surface the mean wind is no longer accelerating, and the stress has become independent of height. The gradient of this surface is about 1/100. The magnitude of this gradient places a stern requirement on any site that is to be used for making observations of equilibrium wind conditions: there must be no significant variation of the roughness of the upwind surface for a distance of at least 100 times the height of the observations.

Transitions from smooth to rough surfaces can be discussed in the same manner.

7.2 Non-stationary Boundary Layers

Over land surfaces the daily heating and cooling cycle causes time rates of change that significantly alter the flows we have been studying up to now. Generally the constant flux approximation is still acceptable in the surface layer, and the Monin–Obukhov profiles can still be expected to prevail. In the Ekman layer, the terms $\partial \overline{u}/\partial t$ and $\partial \overline{v}/\partial t$ cannot be neglected in comparison to the terms that contain the rates of change of stress components with height (see (6.1)).

An understanding of the phenomena that take place in this layer involves several disciplines in addition to turbulence and meteorology. We shall point out some of these in the following discussion, but it is not possible to go into detail in this presentation. Instead, a comprehensive boundary layer model simulation incorporating these features in detail is contained in the diskette that accompanies this publication. By using it as an experimental tool the reader can judge the sensitivity of the atmosphere to various features of the environment.

7.3 The Surface Heat Balance Equation

We now consider what is sometimes called the *active surface*, in which the solar radiation or insolation is intercepted. In the simplest case this is the ground surface. In more complex situations it may be made up of the ground and surfaces of vegetation. In the case of the ocean we consider it to be the ocean surface itself, even though some of the radiation may actually pass through it and be absorbed in the water below. Such exceptions can be dealt with by modifying the definitions of some of the fluxes discussed below.

The importance of this surface is that it is the boundary of the atmosphere, and a knowledge of its properties is necessary for stating the boundary condition of the changing state of the atmosphere. Since we deal with it as a surface, there can be no storage of heat within it; the sum of all the heats that are intercepted must be balanced by the sum of all the heat fluxes that leave it. The equation that states this balance is called the *surface heat balance equation*.

In symbolic form this equation may be written

$$(Q + q)(1 - A) + I_{\downarrow} - I_{\uparrow} = G_0 + H_0 + LE_0 \tag{7.1}$$

The symbols used in this equation are defined and discussed below.

Q Intensity of the direct beam of solar radiation reaching the ground. (Intensity is measured per unit area of the surface, assumed horizontal.)

q Intensity of scattered solar radiation reaching the ground.

A Albedo of the surface.

$I_\downarrow$ Infrared radiative intensity reaching the ground from the atmosphere.

$I_\uparrow$ Infrared intensity emitted upward from the active surface.

G_0 Downward flux of heat into the ground or ocean; this is the rate heat is stored in the ground or ocean. G may be either positive or negative.

H_0 Sensible heat flux to the atmosphere. It is the bounding value of the turbulent heat flux $H = c_p \rho \overline{w'T'}$.

L Latent heat of vaporization ($L \cong 2.5 \times 10^6 \, \mathrm{J\,kg^{-1}}$).

E_0 Water vapor flux leaving the surface (combined evaporation from the surface and transpiration from vegetation).

The combined terms on the left side of (7.1) are usually called the *net radiation*. The solar radiation reaching the ground during the daytime in clear weather can be estimated from the equation

$$(Q + q) = S \cos(\zeta) \left(0.36 + 0.74\tau^{\sec(\varsigma)} \right) , \tag{7.2}$$

in which ζ is the zenith angle of the sun, τ is the *transmissivity* of the atmosphere, and S is the solar constant, equal to about $1360 \, \mathrm{W\,m^2}$. The sun's zenith angle is easily calculated – using astronomical procedures – from the local time, the declination of the sun, and the geographical location of the site. The transmissivity of the atmosphere is the fraction of vertically incident short-wave radiation entering the top of the atmosphere which is transmitted to the ground. It may be estimated from the horizontal visibility V in kilometers by the empirical equation

$$\tau = \exp\left(-9.0/V\right) . \tag{7.3}$$

The downcoming infrared radiation $I_\downarrow$ reaching the ground depends on the distributions of water vapor and temperature in the atmosphere. The estimation of this quantity can be made using a radiation chart, or calculated using two-stream models that are available. There are also empirical equations applicable during clear skies. One of the best of these was devised by Swinbank (1964) (in units of $\mathrm{W\,m^{-2}}$).

$$I_\downarrow = 5.31 \times 10^{-13}T^6 , \tag{7.4}$$

$I_\downarrow$ is often referred to as the *nocturnal radiation*. It should be remembered that it is also present in the daytime, and is then actually slightly greater than at night. When low clouds are present, $I_\downarrow$ is typically larger than with clear skies. The other infrared flux $I_\uparrow$ is easily calculated from the ground temperature using the Stefan–Boltzmann equation

$$I_\uparrow = \varepsilon \sigma T_g^4 \tag{7.5}$$

in which T_g is the absolute ground temperature, ε is the *emissivity* of the surface, and the Stefan–Boltzmann constant σ has the value $5.67 \times 10^{-8} \, \mathrm{W \, m^{-2} \, K^{-4}}$. The emissivity of most surfaces can be assumed to be unity. Both $I_\uparrow$ and $I_\downarrow$ vary relatively little during the 24 hour cycle. The net infrared radiation is typically about $-80 \, \mathrm{W \, m^{-2}}$.

The heat flux into the ground is sensitive to the thermal conductivity and the specific heat of the soil. The value of this flux at the surface is proportional to the vertical gradient of temperature at the surface. Thus the estimation of the soil heat flux usually requires some model for the changing temperature distribution within the soil. The equations governing these changes are (4.16) to (4.18). Anlaytical as well as computational models have been used for this purpose, and these show that when the temperature of the surface undergoes a sinusoidal variation over the diurnal cycle, the ground heat flux G_0 has its maximum about 3 hours before the time of maximum surface ground temperature, or typically about 9 to 10 a.m. local time. In the ocean, G_0 is mainly driven by temperature advection in the atmosphere and the resulting air-sea temperature difference. In the tropics and elsewhere when the advection is weak, sunlight may penetrate sufficiently into the surface layers to influence the stability and surface temperature.

The remaining two fluxes are the sensible and latent heat fluxes. Since they change little with height in the surface layer, it does not matter much whether we discuss the turbulent fluxes or those at the actual interface. One important use of the surface heat balance equation is to determine the sum of the two turbulent fluxes by calculating the remaining terms. In order to separate them further, it is then necessary to know their ratio, which is called the *Bowen ratio*,

$$B \equiv \frac{H_0}{LE_0} \, . \tag{7.6}$$

This ratio can be quite well estimated from ordinary ship observations over the ocean, since the exchange coefficients for water vapor and heat transfer seem to be equal. The ratio is then easily found to be

$$B = \frac{c_p}{L} \frac{\left(\overline{\theta} - \theta_a\right)}{\left(\overline{q} - q_a\right)} \, , \tag{7.7}$$

in which θ_a is the temperature of the sea surface, q_a is the saturation specific humidity at that temperature, corrected for salinity, and $\overline{\theta}$ and $\overline{q}$ are the observed values in the air measured at any selected height in the atmospheric surface layer.

Estimating the evaporation rate over land is more difficult. When bare soil gets hot at the upper surface in the daytime, the upper surface layer quickly dries out. In addition to losing its water to the atmosphere, some water tends to migrate downward by vapor transfer to lower cooler layers. Thus evaporation from the surface tends to be greatest in the morning hours. Vegetation, on the other hand, extracts water from deeper layers in the soil and transfers it to the air through the leaves. This process is controlled effectively by the stomata that open and close in response to the intensity of solar radiation and the availability of water in the soil. Although the processes are complex and involve other disciplines, they are important for the surface layer heat budget. Usually one-quarter to one-half of the heat received at the surface is represented by the latent heat term.

Only when all of the above heat fluxes have been accurately estimated is it possible to use the surface heat balance equation as a boundary condition for the determination of the sensible heat flux H_0. Since the atmosphere gets nearly all of its heat from the surface, it is essential that this process be done correctly if one expects to make useful forecasts of maximum and minimum temperature, the depth of the mixed layer, or other meteorological phenomena that depend on these quantities. It is regrettable that standard meteorological observations fail to give information about the temperature and moisture content of the soil, as models show that the terms of the surface heat balance are quite sensitive to these properties.

The typical diurnal course of terms in the surface heat balance equation over prairies of the midwestern United States is shown in Fig. 7.3, which was prepared from observations made at O'Neill, Nebraska in 1953. It should be noted that 1 Langley min^{-1} is 1 cal. cm^{-2} min^{-1} and is equal to 697.5 W m^{-2}, or about one-half the solar constant.

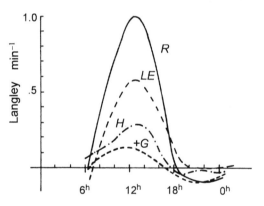

Fig. 7.3 Diurnal course of terms of the surface heat balance equation as observed at O'Neill, Nebraska, August 13–14, 1953

7.4 Daytime Conditions in the PBL

There is usually a capping inversion during the daytime that limits the depth of convection. In the early part of the day, the limiting inversion is the remnant of the previous night's nocturnal inversion. When this disappears, the convective layer grows rapidly until it reaches the subsidence inversion, which is normally found at a height of between one and two kilometers in fair weather. Usually the surface temperature increases rapidly while the nocturnal inversion is present because of

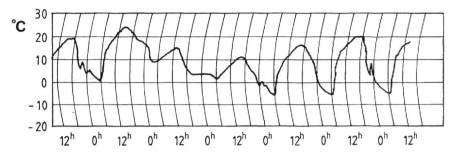

Fig. 7.4 Thermogram recorded in the Barrens of Centre County, Pennsylvania, near State College. All times are local

the shallow depth, to which the heat is confined. Later as the convection deepens, the temperature rise is retarded even though the rate of sensible heating at the surface is greater. These features are seen in Fig. 7.4, which shows an example of the diurnal course of temperature observed at a continental location.

Convection in the daytime is of two kinds, forced convection and free convection. The first is dominated by relatively strong winds and/or weak values of H_0 at the surface. Turbulent energy is generated principally by mechanical production. The eddies tend to be small, and the conditions for K-type mixing are satisfied. Fluxes are determined by local gradients, and mixing is mainly between adjacent layers.

Free convection is characteristic of light winds and strong heating. The surface layer then becomes potentially warmer than the overlying layers, and thermals rise buoyantly out of this layer and continue upward as long as their density is less than the environment. As they rise, the thermals exchange their properties with the surrounding air through entrainment. Thus the transfer is mainly between the surface layer and each of the upper layers rather than between adjacent layers. Thermals continue to rise until they encounter the capping inversion. Usually there is sufficient kinetic energy to enable them to penetrate slightly into the stable region above the inversion, and in doing so, entrain some of the air above the top of the inversion down into the mixed layer. In this way, the inversion is pushed upward, and the mixed layer is deepened.

Except for the heated surface layer, the lapse rate in the convective layer tends to be approximately adiabatic. With forced convection, the potential temperature must decrease with height, at least to a small degree in order for the heat to be transported upward. With free convection, on the other hand, the only requirement is that the surface layer be potentially warmer than the layer above. It is commonly observed that in such layers, the potential temperature increases slowly with height above the immediate vicinity of the surface.

The equation governing the change of potential temperature in the convective layer is

$$c_p \bar{\rho} \frac{\partial \bar{\theta}}{\partial t} = -\frac{\partial H}{\partial z} \tag{7.8}$$

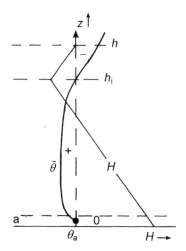

Fig. 7.5 Potential temperature and heat flux distributions during free convection

for a horizontally homogeneous environment. Since $\bar{\theta}$ is approximately independent of height, the temperature changes at about the same rate at every level,
and the heat flux must be an approximately linear function of height. The typical
distribution of H and $\bar{\theta}$ are indicated in Fig. 7.5.

Under free convection conditions, u_* ceases to be a good indicator of the
turbulent velocities. Under very light wind conditions it may approach zero. A
different velocity scale that takes into account vertical velocities generated by
convection can be defined by

$$w_* \equiv \left(\frac{ghH_a}{c_p \bar{\rho}\, \bar{\theta}_a} \right)^{\frac{1}{3}} \tag{7.9}$$

where h is the depth of the mixed layer, g is gravity, and the subscript a refers to
the surface layer. It can be easily shown that

$$\frac{w_*}{u_*} = \left(\frac{h}{-kL} \right)^{\frac{1}{3}} , \tag{7.10}$$

where L is the Monin length. We can characterize forced convection by small
values of this ratio, and free convection by large values. A representative measure
of the turbulent velocities is best characterized by a combination of u_* and w_* of
the form

$$w_m^3 = \left(w_*^3 + 25u_*^3 \right) , \tag{7.11}$$

suggesting that the transition to free convection may occur when the ratio in (7.10)
is as small as 3.

7.5 The Planetary Boundary Layer at Night

Because the outgoing infrared radiation exceeds the incoming rate, the net radiation becomes negative before sunset, sometimes by as much as an hour. By the time the sun actually sets, there is usually already a shallow inversion. The temperature, which starts to fall from the time of maximum in midafternoon, usually falls at its fastest rate around sunset. Later in the night, for reasons that will be discussed later, the rate of decline is smaller and may even be reversed for short periods. These features can be seen by inspecting Fig. 7.4.

In the surface layer the wind is faster in the daytime than at night, but above a few tens of meters, the reverse is true. As a result, the wind shear often becomes very strong in the surface layers, decreasing the Richardson number and promoting conditions favorable for the production of turbulence. When this occurs, heat and other properties are transported downward, and the height of the inversion is increased as a result. Calculations indicate that if radiative transfer acted alone, the layer of nocturnal cooling would extend at most a few tens of meters above the ground. Instead we typically find it extending up to several hundred meters – a sure indication of the importance of turbulence in nocturnal inversions.

During the period from 1916 to 1918, the U.S. Weather Bureau carried out a large experimental kite-flying program in the hope that such platforms would be found suitable for aerological observations. The kites looked a good deal like the airplanes of that era and were of a comparable size. By using multiple kite arrays to support the enormous lengths of steel tethering wires, heights of up to 5 km were reached. The kites were ideal for making slow repeated ascents and descents at night, and dozens of detailed night-time soundings were recorded. On one occasion the wind speed at 180 m was nearly $40 \, \mathrm{m \, s^{-1}}$ while at the ground it was less than $3 \, \mathrm{m \, s^{-1}}$. Descriptions of this program and the observed data make up the first 18 supplements of the *Monthly Weather Review*.

Much of our present understanding about the nocturnal boundary layer came out of a study of the kite data (Blackadar (1957)) and the results of the O'Neill field observing program (Lettau and Davidson (1956)). An important feature is the growth of a wind-speed maximum at the height of the top of the temperature inversion. Although the feature had been observed earlier by Gifford (1952), its regularity and frequency of occurrence had never before been recognized. Gifford's observations are shown in Fig. 7.6.

Several characteristics of low-level jets have been well documented and give clues to the explanation of the phenomenon. The first clue is that the speed is supergeostrophic by an amount that is similar to the amount of the subgeostrophic defect on the preceding afternoon. This fact can be seen in Fig. 7.7. A second characteristic emerged from an analysis of the kite observations: the height of the wind maximum is usually exactly coincident with the top of the layer of nocturnal cooling, as can be seen in Fig. 7.8.

These clues are helpful for understanding the nature of the processes that create the jet profile, as well as the timewise evolution of the cooling at the ground. In

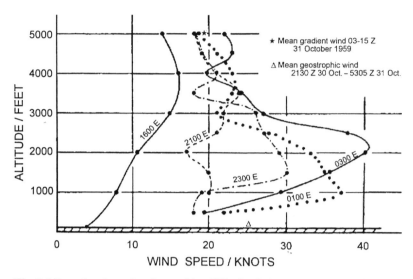

Fig. 7.6 Boundary layer jet observed by Gifford (1952) at Silver Hill, Md, October 30–31, 1950

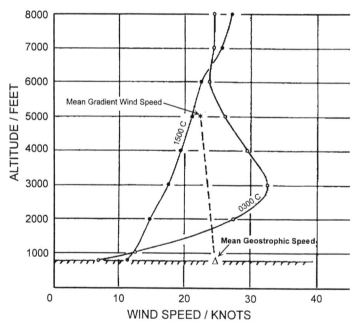

Fig. 7.7 Average wind speed profile for 16 significant boundary layer jets at San Antonio, Texas during January 1953, from Blackadar (1957)

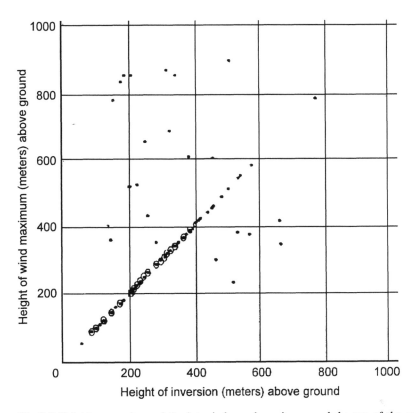

Fig. 7.8 Height comparison of the jet wind-speed maximum and the top of the nocturnal inversion, for 88 selected cases observed from kites between 1916 and 1918 at Drexel, Nebraska; from Blackadar (1957)

the simplest situations we shall assume that the pressure gradient does not change with time. At about the time of sunset, the winds aloft are weak, and a strong shallow inversion develops above the ground. The Richardson number in this inversion becomes quite large, and turbulence is suppressed or eliminated. Thus, rather abruptly, the upper layers which had been strongly retarded by the daytime convection find themselves decoupled from the ground and free to move with little or no friction. The result is the initiation of an inertia oscillation at all levels that had previously been retarded in the afternoon.

In Fig. 7.9, W_0 represents the velocity defect vector at about the time of sunset. With the removal of the surface friction, and assuming that the geostrophic wind is not changing, this defect vector begins rotating to the right (in the northern hemisphere) at an angular rate equal to the Coriolis parameter. For this phenomenon, time is most conveniently measured in pendulum days or pendulum hours, defined as one day, or hour, divided by the sine of the latitude. The maximum wind speed is seen to occur about one-quarter pendulum day after sunset provided turbulent mixing has not influenced its motion up to that time. At 45° latitude, this time is

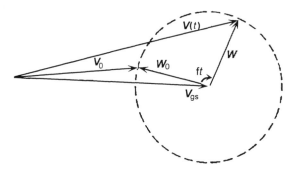

Fig. 7.9 Relation between the complex number W and the wind vector $V(t)$ and the initial values W_0 and V_0 and the geostrophic wind vector V_g during a frictionally initiated inertia oscillation

about nine hours after sunset. If the geostrophic wind is independent of height, the maximum wind speed occurs at the height that was most strongly retarded the previous day. This usually means the lowest level that has not previously become turbulent during the night.

The low level jets in the midwestern United States are particularly spectacular when the general flow is from south to north. It seems very likely that in this area there are other influences that contribute to the large diurnal variation of the wind in the upper boundary layer. There is some indication that the large-scale slope of the plains (about 1/600) though small, is related to a diurnally varying pressure gradient. Actual wind hodographs are ellipses, rather than circles which would characterize simple inertia oscillations.

As the wind speed increases above the surface inversion layer, the wind shear within the inversion increases, and usually becomes sufficient to cause the Richardson number to fall below its critical value. Turbulence then sets in, and heat is brought down to the surface, slowing the rate of cooling, or in extreme cases, reversing it. The same process also brings down momentum, decreasing the wind shear and limiting the further development of turbulence. Thus the nocturnal layer tends to be kept in a state of slight, possibly intermittent turbulence, which causes the layer to slowly increase in depth. The coincidence of the jet with the top of the nocturnal inversion comes about because the turbulence that causes the growth of the temperature inversion is the same as that which removes momentum from the inertially driven layer above.

It is possible to find an approximate relation between the height h of the nocturnal inversion and the distribution of temperature and wind. We make the assumption that at each level within the nocturnal inversion, the Richardson number is maintained at about its critical value. If we make the simplifying assumption that all mean quantities are linear functions of height, we can evaluate the Richardson number from the potential temperatures and wind speed at the ground and the top of the inversion. The result is

$$R_c = \frac{g}{\overline{\theta}} \frac{\left(\overline{\theta}_h - \theta_a\right)/h}{\left(V_h/h\right)^2} = \frac{gh}{\overline{\theta}} \frac{\left(\overline{\theta}_h - \theta_a\right)}{V_h^2} \tag{7.12}$$

or

$$h \cong \frac{0.25\overline{\theta}V_h^2}{g\left(\overline{\theta}_h - \theta_a\right)} . \tag{7.13}$$

If the potential temperature increases by 10° from the ground up to the top of the inversion and the wind speed there is $20\,\mathrm{m\,s}^{-1}$, then the inversion should reach up to about 300 meters.

There are two time scales in the nocturnal layer development. The first is determined by the rate of fall of temperature in the early evening and is related to the thermal characteristics of the underlying surface. Surfaces marked with low conductivity and small heat capacity cool rapidly and favor strong shallow inversions soon after sunset. The second time scale is the reciprocal of the Coriolis parameter, which is set by the rotation of the earth, and is measured in hours rather than in minutes. The numerator of the Richardson number is controlled by the shorter time scale and dominates its behavior early in the night. The second time scale controls the development of the wind shear, which occurs in the denominator of the Richardson number and eventually dominates because it is squared. Thus turbulence is more prominent during the latter part of the night, causing the rate of fall of surface temperature to become slower. When turbulence is intermittent, there may be episodes in which the surface temperature actually increases for a while before resuming its normal decrease. Several examples of such episodes can be seen in Fig. 7.4.

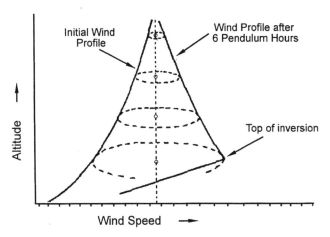

Fig. 7.10 Schematic illustration of the evolution of an inertially induced boundary layer jet profile; from Blackadar (1957)

7.6 Model Simulation of the PBL

Included with this publication is a diskette containing a time-dependent one-dimensional model of the planetary boundary layer suitable for homogeneous locations over continents. All of the programs and accompanying files on this diskette were written by the author.

The model will run on any PC equipped with a graphics color display and Windows 3.1 or Windows 95. It may also run on some monochrome displays and laptops if they are equipped with an internal adapter that emulates the above color displays. The speed of execution is an important factor in the productive use of the program, and if at all possible, it should be run on a machine that is equipped with a math coprocessor.

A description of the model and directions for installing it will be found in Appendix D. The main feature of the program is the display of 22 different default parameters or options, any of which can be accepted by the user as displayed or changed by moving the cursor and typing a new value. The significance of each of these parameters together with other useful information can be displayed by clicking HELP with the mouse after the panel displaying the parameters is chosen from the main menu. It should be noted that any change that is made in the input parameters or options must be terminated with the RETURN or ENTER key; otherwise even though new values appear on the screen, they will not be written into the input data assembly.

The model simulation is intended to be used as an experimental tool. The user is encouraged to run each data set several times. The first run should be made without changing any of the default parameters, which are determined by the observed input data. On each subsequent run it is best to change only one parameter from the original set and thus see most easily the effect that this parameter has on the result. In this way the user can obtain a familiarity with the complex system that would never be possible with the real atmosphere. One can never control the actual atmosphere as one can a simulation or a laboratory experiment.

The first step in running the model is to select one of the data sets. The list is displayed by clicking *Select Case* in the *File* menu. The files referred to in the following discussion are *Oneil3.04d*, *Giffrd.21d*, and *Bogusw.06d*.

When the program begins to run, the display will show the time-changing profiles of wind speed, temperature, and dew point up to 2 km. The model runs in two-minute time steps, and the display may be set to change at every time step or at every fifth time step. The values of each variable are displayed at 100 meter intervals beginning at 60 m, the points at these levels being connected by straight lines. The surface values at a nominal height of 5 meters are displayed at the first pixel above the abscissa, while the values at the actual surface are displayed at the end of the colored line on the abscissa itself. During the middle of the day on the O'Neill example, the surface temperature (average of the ground and grass leaf temperature) is seen to be several degrees warmer than the air temperature, while the temperature at 60 meters is another 2 or 3 degrees colder. A background of blue lines is provided to indicate the dry adiabats.

During the night and early morning hours the model runs in the forced convection mode. The change to free convection is indicated by a set of vertical wavy lines in yellow and a blue symbol $\overline{\Lambda}$ that indicates the height of penetration of the thermals.

An inset in the upper right portion of the display shows the development of the terms of the surface heat balance equation. The terms are identified by color as explained in the display. Note that the sum of G_0, H_0, and LE_0 is always equal to the driving function, the net radiation. An icon indicates when the sun is above the horizon and its approximate elevation in the sky as the day progresses. The local time is also shown.

Whenever a layer contains liquid water, an ###### icon is displayed at the appropriate height. If the surface layer has liquid water, the symbol ====== is displayed instead. Layers with liquid water are always treated as black bodies when the infrared radiative transfer option (i.e., the default option) is activated.

A more complete discussion of the architecture of the model and its operation is given in Appendix D. In the remainder of this chapter some notes and interesting features will be discussed for each of the three input data sets referred to above.

The first example, *Oneil3.04d*, makes use of a data set observed at O'Neill in August 1953. It begins at 4 a.m. local time, at which time the wind speed profile shows a low-level jet and a strong nocturnal inversion. As the sun comes up, the wind and temperature undergo typical transitions to the daytime free convection. With the default subsidence rate, the thermals rise in excess of 2 km. Notice how the nocturnal inversion begins to form in the late afternoon about half an hour before sunset, and the rapid drop of surface temperature at about the time of sunset. One particularly interesting experiment is to increase the subsidence rate from the default value of $1 \, \text{cm s}^{-1}$ to $3 \, \text{cm s}^{-1}$. The effect of subsidence is easy to see, since its total for an hour is injected all at once each hour on the hour and displayed at the first change of display after the hour begins. Notice how the previously existing subsidence inversion descends from above 2 km initially to about 1 kilometer before it is wiped out by the penetration of the thermals, which by the end of the afternoon produce a new inversion at a higher level. Experiment also with changes of soil moisture and see the effect of these changes in the maximum and subsequent minimum temperatures.

The second example, *Giffrd.21d*, is Gifford's classic low-level jet observation, which was accompanied by episodes of temperature rise at the surface during the course of the night. The run begins at 2100 local time when a strong temperature inversion is already present. Watch the development of the low-level jet and the gradual increase of wind shear beneath it. At about midnight the downward flux of heat becomes so strong that the surface temperature increases for a while. The downward flow of heat is accompanied by a downward flow of momentum, and the result is an upward progression of both the top of the temperature inversion and also the wind speed maximum. Try repeating the run with different geostrophic wind speeds and different latitudes. Note also, particularly at higher latitudes that the jet begins to diminish long before sunrise, showing that it is an inertial phenomenon rather than a simple response to the heating and cooling cycle.

The third example, *Bogusw.06d*, is a bogused data set originally observed in the Beauce district of central France. Since good data sets tend to be made in anticyclonic weather situations, few have sufficient water vapor to give stratus and stratocumulus clouds below two kilometers. For this reason, additional water vapor was provided to the input data artificially. Clouds form almost immediately between one and two kilometers and also in the lowest two layers. The sun climbs in the sky but little radiation reaches the ground because of the cloud cover. Thus the surface layer inversion persists throughout the entire day. Once each hour, subsidence injects dry air into the cloud layers, but the cloud layers remain intact. Careful study of the developing profiles of temperature and dew point show all the indications of a well mixed layer above the nocturnal inversion: the lapse rate is adiabatic, dry-adiabatic beneath the clouds and moist adiabatic in the cloud layers, while the dew-point lapse rate is that which accompanies a constant mixing ratio below the clouds. Clearly the instability is not a result of heating from below. In fact it is caused by infrared radiative cooling from the top of the cloud layer. In spite of the subsidence which would normally heat the cloud layers, the layers are slowly cooling and a strong inversion forms above the cloud top. Persistent stratus cloudiness of this kind is very common over Europe and the western United Status in the winter. That such clouds are caused by radiation may be verified by repeating the run with the infrared radiative transfer turned off. (This option does not affect the infrared terms in the surface heat balance equation, but it does prevent radiative cooling from affecting layers above the surface.) It will be seen that without radiative cooling, the cloud disappears within a few hours.

7.7 Problems

1. Prove (7.7) assuming that $K_q = K_m$ and that the fluxes are independent of height up to the level of measurement.

2. After installing the diskettes, run B.L.Model and select the case Gifford.21d. When the first panel of parameters is displayed, change the latitude from 39° to 0°. Be sure to press the ENTER key after you make the change so it will be permanently entered into the input array. Why does no low-level jet form in this case? Why does fog form at this low latitude when it did not form at 39°? Continue the run until the fog disappears and note the time.

3. Repeat the above run. In addition to the latitude change, disable the infrared cooling. Explain any difference observed between this run and the one in #2 above.

4. Look at Fig. 7.1. Will the stress under the surface AB1 be larger than that above AB2 or smaller? Why? Following the mean motion of the air between the two surfaces, is the speed increasing or decreasing?

5. Redraw Fig. 7.2 to show the wind profiles that would prevail when the wind moves from a smooth surface onto a rough one.

6. If (7.2) were completely correct we would never be able to see the sun set. Obviously something has been ignored in deriving it. What is it, and how large are the errors likely to be?

7. People sometimes mistakenly use the surface air temperature to estimate the outgoing long-wave radiation from the surface. If the ground temperature is 10°C warmer than the air temperature in the daytime, how large would the error be in absolute units and in percentage units?

8. The Barrens of Central Pennsylvania (see Fig. 7.4) are noted for cold temperatures at night. For example, on the next to last day in the illustration, the temperature climbed 26°C in 10 hours from the night-time minimum to the day-time maximum. This large range is due in part to the nature of the surface and in part to the topography of the area. With these facts in mind, give your appraisal of the probable nature of this area.

9. Show that during a period of free convection, the total rate of TKE production in the mixed layer is given approximately by one half the rate of production at the surface times the depth of the mixed layer h. Assume h is the same as the level h_i shown in Fig. 7.5. Also, neglect the energy produced within the surface layer.

10. The dew point is indicative of the total water-vapor content of the air. At typical continental locations it varies significantly over a 24-hour cycle, although the magnitude of the variation is not large. This variation has a double maximum and minimum as shown be the following averages for Kiev during June, July, and August 1973–1990.

Local time hr	Dew point °C	Local time hr	Dew point °C
02	12.23	14	12.22
05	11.85	17	11.50
08	11.50	20	11.56
11	13.17	23	12.67

Discuss the boundary layer processes that are involved and explain the variations.

8 Statistical Representation of Turbulence I

It is important that we look in more detail at the characteristics of turbulence and its distribution in the atmosphere under various meteorological conditions. This information will have important applications in the construction of better models of mean flow and in estimating the diffusion of properties emitted from chimneys, automobiles, and other sources.

A number of important statistical characteristics of turbulence can be defined by means of the *probability density* of a function $u(t)$. We have used $u(t)$ to refer to the x-component of the velocity. However, we can define a probability density function for any fluctuating property. We shall refer to the probability density function as $\beta(u, t)$. The probability that $u(t)$ falls in the range between u and $u + du$ is by definition $\beta(u, t)du$, and from this definition it follows

$$\int_{-\infty}^{+\infty} \beta(u, t)\, du = 1 \,. \tag{8.1}$$

We define the nth *moment* of the distribution $\beta(u, t)$ as

$$\overline{u^n} = \int_{-\infty}^{+\infty} u^n(t)\, \beta(u, t)\, du \,. \tag{8.2}$$

The first moment is $\overline{u}(t)$, and we shall assume for now that this is zero. The second moment is the *variance* $\overline{u^2}(t)$. The third moment is called the *skewness*, and the fourth moment is called the *kurtosis*. If the distribution is symmetric, the odd moments are all zero. *Stationarity* implies that the probability is independent of time t.

In the same way we can define the *joint probability* density $\beta_2(u_1, u_2, t_1, t_2)$ of two variables $u_1(t)$ and $u_2(t)$. $\beta_2(u_1, u_2, t_1, t_2)du_1 du_2$ is the probability that the product of the two variables lies in the rectangle bounded by $u_1,, u_2, u_1 + du_1$ and $u_2 + du_2$

$$\int_{-\infty}^{+\infty} \int_{-\infty}^{+\infty} \beta_2(u_1, u_2, t_1, t_2)\, du_1 du_2 = 1 \,. \tag{8.3}$$

As above we define *joint moments*

$$\overline{u_1^m(t_1)u_2^n(t_2)} = \int_{-\infty}^{+\infty} \int_{-\infty}^{+\infty} u_1^m u_2^n \beta_2(u_1, u_2, t_1, t_2)\, du_1 du_2 \,. \tag{8.4}$$

The simplest of these, and the one most often of interest, is the *covariance*, defined for $t_1 = t_2 = t$.

$$\overline{u_1 u_2}(t) = \int_{-\infty}^{+\infty} \int_{-\infty}^{+\infty} u_1(t) u_2(t) \beta_2(u_1, u_2, t) du_1 du_2 . \tag{8.5}$$

Most joint statistics are functions of the time separation between t_1 and t_2. Statistics that are functions only of $\tau = t_2 - t_1$, and therefore not functions of t_1 or t_2 separately, are *stationary*.

The *autocorrelation* comes from the special case whre u_1 and u_2 are values of the same function at two different times t and $t + \tau$.

$$\overline{u(t_1) u(t_2)} = \overline{u^2} \rho(\tau) , \quad \tau = t_2 - t_1 . \tag{8.6}$$

The function on the left is the *autocovariance*, and $\rho(\tau)$ is the *autocorrelation*. Since t_1 and t_2 can be interchanged, ρ is an *even* function of τ. That is

$$\rho(-\tau) = \rho(\tau) . \tag{8.7}$$

It is obvious that when $\tau = 0$, $\rho(\tau) = 1$. It is generally assumed that as $\tau \to \infty$, $\rho(\tau) \to 0$. This cannot be proved in the atmosphere because real statistics are not completely stationary. The *integral time scale* is defined as

$$T = \int_0^{\infty} \rho(\tau) d\tau \tag{8.8}$$

if it exists. Its existence is usually assumed, but in practice it is necessary to estimate it from records of finite duration. A little thought evokes a multitude of problems. Infinite time encompasses many synoptic situations each of which must be expected to have its own turbulence characteristics. The integral time scale during a thunderstorm must be different from what prevails during the passage of a large anticyclone. Need one say more?

Statistics can be of two basically different kinds. *Eulerian* statistics refer to functions measured at a fixed location while *Lagrangian* statistics refer to functions measured on a moving air particle. Most important applications of turbulence have to do with transport and diffusion of properties and involve Lagrangian statistics. But most of the available measurements are made at a fixed location for obvious reasons. One method of obtaining Lagrangian statistics is through the use of *tracers*: identifiable materials that mix or become suspended in the air and which have a known origin in space and time. The information provided by such measurements is, however, quite limited, and knowledge of how to predict transport and diffusion is greatly hindered by the difficulty of getting good Lagrangian statistics.

8.1 Scaling Statistical Variables in the PBL

A number of scaling principles have been already considered. In this section, we review these and consider two other schemes that have been considered for generalizing the distribution of turbulent statistics.

In the surface layer, mechanical energy production is nearly always a significant process, especially very close to the surface. The mean velocity and the vertical component of the turbulent velocities scale as u_*, and where mechanical production is dominant, heights scale as z_0. Mean variables and statistics are either independent of height or are distributed logarithmically. Elsewhere in this layer, i.e., where buoyant production plays an important role, the height scales as the Monin length

$$L = - \frac{c_p \overline{\rho T} u_*^3}{kgH} ,$$
(8.9)

the winds with u_* and the temperatures with θ_* defined as $-H/c_p \rho u_*$. The simple logarithmic distributions are replaced by Monin–Obukhov similarity, which includes the logarithmic distribution as a limiting case.

We have arrived at a successful scaling of the neutral planetary boundary layer by scaling the velocity defects by u_*, and height by u_*/f. It should be pointed out, however, that deviations from the geostrophic wind are so small above about 100 meters that it is virtually impossible to measure them. Therefore, the validity of using this length is mostly justified by the lack of any suitable alternative.

Under free convective conditions it becomes possible to use the height of the mixed layer, which we shall denote as h, as a length scale in place of u_*/f. Many statistics in the mixed layer seem to be best described by representing their distribution with reference to the dimensionless height z/h. We have also seen that the convective velocity scale w_*, defined by

$$w_* \equiv \left(\frac{ghH}{c_p \overline{\rho T}} \right)^{\frac{1}{3}} ,$$
(8.10)

seems to be more appropriate than u_* under these conditions. This kind of scaling is usually called *mixed layer scaling*. It should be emphasized that the variables used in defining these scales (H, τ_0, $\overline{T}$, etc.) are the values prevailing at the underlying surface. The transition from neutral to mixed layer scaling probably occurs with the changeover from forced to free convection.

Stable conditions are the most difficult to deal with, partly because turbulence is frequently intermittent or perhaps not be present at all. Nieuwstadt (1984) has pointed out that height is usually not a useful variable for describing the distribution of turbulence statistics in these conditions; the turbulent vortices are small and generally are so far away from the ground that they cannot be influenced by it in any meaningful way. Statistics must then be determined by purely local scales. Such similarity is called *local similarity* or by Nieuwstadt's term *z-less similarity*.

Generally the scales used for local similarity are defined in the same way as those used for Monin–Obukhov similarity, except that the turbulent parameters involved in the definitions are those prevailing locally rather than at the ground. For example a local velocity scale can be defined as

$$u_{*1} \equiv \sqrt{-\overline{w'u'}} \tag{8.11}$$

and from this, a local length scale

$$\Lambda \equiv -\overline{T}u_{*1}^3/kg\overline{w'T'} \ . \tag{8.12}$$

and local temperature and humidity scales

$$\theta_{*1} \equiv -\overline{w'T'}/u_{*1} \ ;$$

$$q_{*1} \equiv -\overline{w'q'}/u_{*1} \ . \tag{8.13}$$

We can note that in the actual surface layer, the parameters that determine the local scales $(-\overline{w'u'}, -\overline{w'T'}$, etc.) are independent of height, and are therefore identical to those defined using surface values. Thus local similarity is not necessarily different from Monin–Obukhov similarity.

Recently Shao and Hacker (1990) have shown that local similarity can also be successfully applied to the unstable planetary boundary layer. As in the case of stable layers, equilibrium conditions imply a relation between the vertical distributions of heat flux and stress, so that local scales can be recast in terms of scales determined from surface values. However, locally defined scales have a

		Summary of Scaling Parameters		
			Mixed Layer	
Dimension	Sfc. Layer	Free	Forced	Local (NBL)
Length	z_0	h	$\dfrac{u_*}{f}$	$\Lambda = -\dfrac{\overline{T}_0 u_{*1}^3}{kg\overline{w'T'}}$
	$L = \dfrac{c_p \overline{\rho}\,\overline{T}_0 u_*^3}{kgH}$		L	
Velocity	$u_* = \left(\dfrac{\tau_0}{\overline{\rho}}\right)^{\frac{1}{2}}$	$w_* = \left(\dfrac{gH_0 h}{c_p \overline{\rho}\,\overline{T}_0}\right)^{\frac{1}{3}}$	$u_* = \left(\dfrac{\tau_0}{\overline{\rho}}\right)^{\frac{1}{2}}$	$u_{*1} = \left[\left(\overline{w'u'}\right)^2 + \left(\overline{w'v'}\right)^2\right]^{\frac{1}{4}}$
Temperature	$T_* = -\dfrac{H_0}{c_p \overline{\rho} u_*}$	$\theta_* = -\dfrac{H_0}{c_p \overline{\rho} w_*}$	T_*	$T_{*1} = -\dfrac{\overline{w'T'}}{u_{*1}}$
Scalar	$q_* = -\dfrac{\left(\overline{w'q'}\right)_0}{u_*}$	$Q_* = -\dfrac{\overline{w'q'}}{w_*}$	q_*	$q_{*1} = -\dfrac{\overline{w'q'}}{u_{*1}}$

Fig. 8.1 Table of scales found to govern the principal dimensions in the atmospheric boundary layer. Note that subscript 0 refers to values at the surface. Fluxes used in local scaling, indicated by subscript 1, are the prevailing values at the situs of evaluation

wider applicability. For example, the distribution of turbulent statistics in a deep layer that is in transition from one kind of surface to another with a different roughness and rate of heating seems to be well described when expressed in terms of locally defined scales. The problem with locally defined scales is that they are not immediately useful when the variables or statistics have to be predicted in space and time.

It is also possible for Monin–Obukhov and mixed layer scaling to both apply at the same overlapping region, but special conditions must be satisfied.

Figure 8.1 summarizes the scales that are assigned in each type of scaling to each of the dimensions that make up physical quantities.

The following relations between scales are also often useful:

$$\frac{\theta_*}{T_*} = \frac{Q_*}{q_*} = \frac{u_*}{w_*} = \left(-\frac{kL}{h}\right)^{\frac{1}{3}} \tag{8.14}$$

and

$$\frac{z}{-L} = \left(\frac{w_*}{u_*}\right)^3 k\frac{z}{h} \tag{8.15}$$

8.2 Vertical Distributions of the Variances

The number of observations of variances above the surface layer is not very large. To make the maximum use of the available data, we use the scales we have found to be successful to nondimensionalize the variances. Then we apply similarity principles that have been found to be successful in our previous studies in order to assist us in making hypotheses. Finally, we subject these hypotheses to comparison with available data to draw conclusions about relationships that will be useful in future modeling studies.

We begin with the variance of vertical velocity, which we can represent using the standard deviation σ_w. Since it is a velocity, we scale it using u_*. A reasonable hypothesis for the surface layer is that this ratio obeys Monin–Obukhov similarity,

$$\frac{\sigma_w}{u_*} = \varphi_3\left(\frac{z}{L}\right) . \tag{8.16}$$

Wyngaard et al. (1971) found that data from a large field program in Minnesota satisfied the relation

$$\frac{\sigma_w}{u_*} = \left(\frac{6.9z}{-L}\right)^{1/3} \tag{8.17}$$

quite well under lapse conditions. Panofsky et al. (1977) have recommended a slightly different representation for unstable cases;

$$\frac{\sigma_w}{u_*} = 1.25\left(1 - 3\frac{z}{L}\right)^{1/3} . \tag{8.18}$$

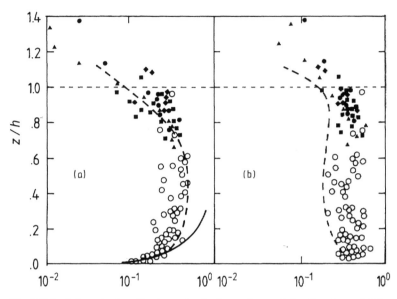

Fig. 8.2 Scaled vertical and horizontal velocity variances [after Caughy and Palmer (1979)]. Solid lines are the free-convection prediction; dashed lines are from Willis and Deardorff (1974)

As we move upward out of the surface layer in lapse conditions, we expect mixed layer scaling to replace Monin–Obukhov scaling. Accordingly we consider the ratio $\sigma_w/w*$ to be the appropriate non-dimensional variable to use, and similarity principles suggest that this ratio should be a function only of z/h, where h is the depth of the mixed layer. One hypothesis for the nature of this function can be made by taking the limit of (8.17) or (8.18) as $z/-L$ becomes very large. Both give about the same result

$$\frac{\sigma_w}{w_*} = 1.33 \left(\frac{z}{h}\right)^{1/3} , \tag{8.19}$$

but this is not supported by observational data (see Fig. 8.2). A reasonably good fit of observational data for the lower half of the mixed layer is the empirical formula

$$\frac{\sigma_w}{w_*} = 0.84 \left(\frac{z}{h}\right)^{0.22} . \tag{8.20}$$

Other relationships have been suggested by Hojstrup (1982) using a theoretical model. The few observations available from above the mixed layer indicate that σ_w probably does not scale with u_* or w_*, and is quite small.

Under stable conditions there is a great deal of scatter in observations. It is generally concluded that σ_w/u_* is independent of z/L and has a value of about 1.3. This result is in agreement of Nieuwstadt's principle of z-less similarity. Use of local scaling suggests that under these conditions,

$$\frac{\sigma_w}{u_{*1}} = f\left(\frac{z}{\Lambda}\right) \tag{8.21}$$

and z-less similarity means that this function must be a constant. Nieuwstadt found it to be 1.4. Note that when the Richardson number exceeds the critical value, all of the local scales go to zero or are indeterminate.

The standard deviations of the horizontal components of velocity do not obey Monin–Obukhov similarity. Under lapse conditions they are strongly dependent on stability and appear to be determined by downward extensions of large eddies from the mixed layer above. Panofsky et al. (1977) have found that σ_u/u_* and σ_v/u_* are well represented in the surface layer by the relation

$$\frac{\sigma_v}{u_*} = \frac{\sigma_u}{u_*} = \varphi_2\left(\frac{z}{-L}\right) = \left(12 - \frac{0.5h}{L}\right)^{1/3} . \tag{8.22}$$

This relation can be converted to mixed layer scaling using (7.10)

$$\frac{\sigma_v}{w_*} = \left(-12\frac{kL}{h} + 0.2\right)^{1/3} \tag{8.23}$$

suggesting that it is independent of z/h. This conclusion is borne out by Fig. 8.2.

The variance of passive scalar quantities such as temperature and water vapor should also satisfy Monin–Obukhov similarity in the surface layer. Thus, if q is used to indicate any such scalar property, then σ_q/q_* should be a universal function of z/L. Under unstable conditions, one should also expect that in the mixed layer, σ_q/Q_* should be a universal function of z/h. If this is the case, the limiting form of the functions can be determined, since there must be a broad overlapping region defined by $z \gg -L$ and $z \ll h$, where σ_q is described identically by functions of either variable. If we define $\zeta \equiv (z/-L)$ and $\xi \equiv (z/h)$, then

$$\frac{\sigma_q}{q_*} = \varphi_Q(\zeta) \quad \text{and} \quad \frac{\sigma_q}{Q_*} = \Phi_Q(\xi) . \tag{8.24}$$

We differentiate both functions with respect to ζ and require that they both be equal over a range of heights. Noting (8.14) and $d\xi/d\zeta = (-L/h)$, one finds

$$\zeta^{4/3}\frac{d\varphi_Q}{d\zeta} \equiv k^{1/3}\xi^{4/3}\frac{d\Phi_Q}{d\xi} . \tag{8.25}$$

Since the left side is a function of ζ only and the right side is a function of ξ only, both must be equal to a constant, which we can define as $\alpha/3$. Integration of the left side then gives the limiting form

$$\frac{\sigma_q}{q_*} = \alpha\left(\frac{z}{-L}\right)^{-1/3} . \tag{8.26}$$

The constant α has been found to be close to 1.0.

The form of this function in the surface layer can be further explored by means of exchange theory. If q is a conservative property, its deviations from the mean

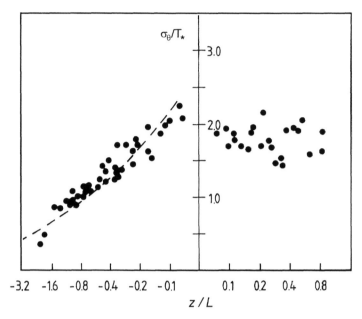

Fig. 8.3 Values of σ_θ/T_* from measurements at the Kansas field program (Wyngaard and Cote (1971)). The dashed line is 3.5 times the nondimensional temperature gradient ϕ_h as approximated by the Businger–Dyer formula. Stable layers do not follow Monin–Obukhov similarity

are associated with displacements l as noted in Fig. 1.3. From (1.21) and (1.23) we obtain

$$\sigma_q^2 = l^2 \left(\frac{\partial \overline{q}}{\partial z}\right)^2 , \tag{8.27}$$

where l is the mixing length. In unstable surface layers, this length is proportional to z. Thus

$$\left|\frac{\sigma_q}{q_*}\right| = a\frac{z}{q_*}\frac{\partial \overline{q}}{\partial z} = \frac{a}{k}\varphi_q\left(\frac{z}{L}\right) \tag{8.28}$$

where a is some undetermined constant and φ_q is the dimensionless gradient of $\overline{q}$. Experience shows this to be well represented by the Businger–Dyer formula for φ_h. The result is

$$\left|\frac{\sigma_q}{q_*}\right| = c\left(1 - 16\frac{z}{L}\right)^{-1/2} \tag{8.29}$$

where c is a new constant. Figure 8.3 shows observations of σ_θ/T_* reported by Wyngaard et al. from the Kansas field program together with the predicted form using c = 3.5. Others have reported values of c ranging from 2 to 5. It should be remembered that the Businger–Dyer formula and therefore (8.29) do not have the correct asymptotic behavior for large $z/-L$ as required by (8.26).

8.3 Problems

1. Show that the correlation coefficient between the vertical velocity and a scalar quantity such as water vapor obeys Monin–Obukhov similarity in the unstable surface layers. Assuming c defined in (8.29) has the value 5, what is the limiting value of the correlation coefficient r close to the surface?

2. We shall see later that one of the most important statistics for predicting horizontal dispersion of a smoke plume is the standard deviation of the horizontal wind directions σ_α. For small angles, this statistic is given by $\sigma_\alpha \cong \sigma_v/\overline{u}$, where v is the component of the velocity at right angles to the mean wind direction and $\overline{u}$ is the mean wind speed. Find an equation for this statistic as function of height in the unstable surface layer.

3. From the result of the last problem, discuss the effect of various conditions of wind speed and surface heating on σ_α.

4. Write an expression that represents, to a first approximation, the relation between the covariance of vertical velocity and temperature as a function of z/h in the mixed layer under free-convective conditions. (See Sects. 7.8–7.9.) Does this statistic satisfy Monin–Obukhov similarity? Does it satisfy mixed-layer similarity?

9 Statistical Representation of Turbulence II

9.1 Spectrum and Cross Spectrum of Turbulence

The spectrum of some function u of a turbulent environment reveals how the variance of this function is distributed among the many frequencies that characterize its variations in time or space.

We begin by considering temporal frequencies and observations of some function u over a period of time. The aim is to describe the distribution of the variance of u over the full range of time-frequencies from the longest to the shortest. The angular frequency ω is defined as 2π divided by the time period of the repetition of the particular frequency. We define the spectrum as the Fourier transform of the autocovariance $\rho(\tau)$

$$S(\omega) \equiv \frac{1}{2\pi} \int_{-\infty}^{+\infty} e^{i\omega\tau} \overline{u^2} \rho(\tau) \, d\tau \ . \tag{9.1}$$

If the spectrum is known for all frequencies ω, one can find the autocovariance from the inverse transform

$$\overline{u^2}\rho(\tau) = \int_{-\infty}^{+\infty} e^{-i\omega\tau} S(\omega) \, d\omega \ . \tag{9.2}$$

The spectrum is real; therefore it can be represented by a cosine transform

$$S(\omega) = \frac{\overline{u^2}}{\pi} \int_0^{\infty} \cos(\omega\tau) \rho(\tau) \, d\tau \tag{9.3}$$

and the autocovariance by the inverse transform

$$\overline{u^2}\rho(\tau) = 2 \int_0^{\infty} \cos(\omega\tau) S(\omega) \, d\omega \ . \tag{9.4}$$

We can interpret $S(\omega)d\omega$ as the fraction of the total variance having angular frequencies between ω and $\omega+d\omega$. The spectrum is thus the density of the variance in frequency space and for this reason it is often referred to as the *spectral density*.

One can analyze the distribution of the joint moments of two variables in frequency space in the same way. The simplest of these is the *cross-variance* between two variables u and v:

$$C_{uv}(\tau) = \overline{u(t_1)v(t_2)}; \quad \tau = t_2 - t_1 . \tag{9.5}$$

Proceeding as above, one obtains the *cross-spectrum*

$$S_{uv}(\omega) \equiv \frac{1}{2\pi} \int_{-\infty}^{+\infty} e^{i\omega\tau} C_{uv}(\tau)\, d\tau . \tag{9.6}$$

The inverse transform is

$$C_{uv}(\tau) = \int_{-\infty}^{+\infty} e^{-i\omega\tau} S_{uv}(\omega)\, d\omega . \tag{9.7}$$

A function is said to be *even* if its value at t_* is the same as its value at $-t_*$. An *odd* function is one whose corresponding values have the opposite sign but are otherwise equal. Every function defined in the range from $-\infty$ to $+\infty$ can be represented as the sum of an even function and an odd function. The integral from $-\infty$ to $+\infty$ of an odd function must be zero, and the integral of an even function over the same range is twice the integral from 0 to ∞.

Since $C_{uv}(\tau)$ is not an even function, the cross-spectrum $S_{uv}(\omega)$ has both real and imaginary parts. In this respect it is helpful to divide $C_{uv}(\tau)$ into its even part $E_{uv}(\tau)$ and odd part $O_{uv}(\tau)$:

$$E_{uv}(\tau) = \frac{1}{2}[C_{uv}(\tau) + C_{uv}(-\tau)] \tag{9.8}$$

$$O_{uv}(\tau) = \frac{1}{2}[C_{uv}(\tau) - C_{uv}(-\tau)] \tag{9.9}$$

from which it is seen

$$C_{uv}(\tau) = E_{uv}(\tau) + O_{uv}(\tau) . \tag{9.10}$$

From (9.6) it then follows

$$\begin{aligned} S_{uv}(\omega) &= \frac{1}{2\pi} \int_{-\infty}^{+\infty} E_{uv}(\tau)\cos(\omega\tau)\, d\tau + \frac{i}{2\pi} \int_{-\infty}^{+\infty} O_{uv}(\tau)\sin(\omega\tau)\, d\tau \\ &= Co_{uv}(\omega) + iQ_{uv}(\omega) . \end{aligned} \tag{9.11}$$

Co_{uv} is called the *cospectrum* and Q_{uv} is called the *quadrature spectrum*.

When $\tau = 0$, C_{uv} is the covariance (see (9.5)) and is real ($Q_{uv}(0) = 0$). Thus we interpret the cospectrum as the spectral density of the covariance.

Many physical variables bear a relationship to each other that involves a change of phase: the current and voltage across a capacitance are $90°$ out of phase; the rate of rainfall and the pressure also tend to be $90°$ out of phase; etc. The quadrature spectrum provides a frequency analysis of this kind of relationship. We can look at a special case. We choose any frequency ω and calculate the lag time that causes a change of phase for that frequency by an angle θ. We then make $v(t)$ equal to $u(t)$ except for a phase shift of θ. The spectra of u and v are then equal;

$$S_u(\omega) = S_v(\omega) . \tag{9.12}$$

Because of the angular displacement of the two time functions, one observes

$$Co_{uv}(\omega) = \cos\theta\ S_u(\omega)\ ;\quad Q_{uv}(\omega) = \sin\theta\ S_u(\omega)\ . \tag{9.13}$$

In this way we can interpret the cospectrum as the spectrum of those variations of the two functions that are in phase with each other, while the quadrature spectrum describes the density of the cross variance that is 90° out of phase. At any frequency, the total degree of relationship between the two functions is measured by the *coherence* defined by

$$Coh(\omega) = \frac{Co_{uv}^2(\omega) + Q_{uv}^2(\omega)}{S_u(\omega)\,S_v(\omega)}\ . \tag{9.14}$$

9.2 Spatial Representation of Turbulence

We choose the x-axis parallel to the mean wind and assume there is a function $u(x_1)$ that represents the values of u along the x_1 axis at some instant in time. We define f to be the frequency (number of waves per unit time) observed at a point at a fixed location. The frequency f is related to ω by

$$f = \omega/2\pi\ . \tag{9.15}$$

An important conjecture, known as Taylor's hypothesis, states that the spatial statistics of u can be constructed by assuming that the $u(x_1)$ function is 'frozen in time' and displaced along the x_1-axis at the speed u of the mean wind. Thus if k_1 is the wavenumber (number of waves per unit distance) in the x_1 direction,

$$k_1 = \frac{f}{\overline{u}}\ . \tag{9.16}$$

If we observe the statistics (i.e., the spectrum, cospectrum, etc.) as functions of f, we can then convert them to functions of k_1.

In general the wave number is a vector with three components k_1, k_2, and k_3 in the x_1, x_2, and x_3 directions. The three-dimensional spectrum shows the distribution of u with respect to k, the magnitude of the wavenumber vector. This representation is particularly suitable when the turbulence is isotropic. The spectrum described in the preceding paragraph is a one-dimensional spectrum. Problems arise when the turbulence is not isotropic. In an extreme case, suppose that the variance of the velocity vector consists entirely of waves whose phase surfaces are parallel to the xy-plane. In this case k_3 is k, and both k_1 and k_2 are zero. Thus the one-dimensional spectrum defined above would not show the distribution of u in three-dimensional wavenumber space. In practice we do not worry much about this distinction since turbulence in the boundary layer is reasonably diverse in directionality.

It should be pointed out that an alternative practice is commonly found in the literature to define the wavenumber as the number of radians per unit length rather than the definition used here, viz. the number of cycles per unit length. The effect of this alternative is to replace f in (9.16) by ω, the number of radians per unit time. This alternative may also necessitate a change in the value of certain constants.

9.3 The Equilibrium Theory of Turbulence

Most of our present understanding of turbulence spectra is centered on what is usually called the *equilibrium theory*. This theory was comprehensively constructed by A. N. Kolmogorov (1941), a prominent Soviet academician, during and immediately following World War II. During this period the normal lines of communication between scientists of different countries were almost non-existent. It later became apparent that many of the ideas that are credited rightly to Kolmogorov were independently conceived by others, amongst whom were W. Heisenberg (1948), L. Onsager (1945), and G. F. von Weizsaecker (1948).

Kolmogorov pointed out that the energy containing eddies of the turbulence spectrum are those with the smallest wavenumbers (i.e. the largest in size) and these are usually anisotropic. They display strongly the configuration of the boundaries and the processes that generate their energy. For example, in free-convection, the vertical velocities tend to be the most prominent of the low wavenumber spectra in the mixed layer. The energy-containing eddies are the most effective contributors to turbulent transport.

Turbulent kinetic energy is generated at the lower wavenumbers. However, it is dissipated almost exclusively at very high wavenumbers. If one considers the entire range of wavenumbers that comprise the turbulent energy, then the subrange within which the energy is removed to a significant degree is a very small part at the high wavenumber end.

The dissipation rate ε is proportional to the square of the velocity gradient. Only the eddies with the smallest sizes (largest wavenumbers) have large velocity gradients, and as a result, the dissipation subrange is restricted to the highest wavenumbers of the spectrum. It can be shown that the spectrum of ε is $k^2 S(k)$. Thus the dissipation spectrum is strongly peaked at high wavenumbers. Heisenberg (1948) showed, using a model, that in this subrange,

$$S(k) \propto k^{-7} .$$
(9.17)

The dissipation rate ε depends on the kinematic viscosity ν. From ε and ν one can form a new length by dimensional analysis:

$$\eta \equiv \left(\frac{\nu^3}{\varepsilon} \right)^{1/4} .$$
(9.18)

This length is the Kolmogorov *microscale*, and its reciprocal is the wavenumber that characterizes the part of the spectrum that is dominated by the dissipation process. In the atmospheric surface layer, $\eta \sim 1$ mm. Thus most of the spectrum is not influenced significantly by viscosity.

For most systems, and in particular the atmosphere, the Reynolds number is very large. As we have seen previously, this means that between the wavenumber of the energy-containing eddies and the wavenumbers influenced by viscosity there are many orders of magnitude. The energy put in at the low wavenumber end of the spectrum must find its way to the high wavenumber end. This process, often referred to as the energy cascade is carried out by the process of vortex stretching. As the energy is transformed from the original low wavenumber anisotropic eddies to the higher wavenumbers, they lose their 'memory' of the configuration that produced the energy and they become more and more isotropic. Thus, if the Reynolds number is large, there exists a broad range of wavenumbers, where the turbulence is independent of viscosity. This portion of the spectrum is called the *inertial subrange*. Within most of the inertial subrange, the turbulence is also isotropic, and this range, which also continues into the viscous subrange, is called the range of *local isotropy*.

9.4 The Inertial Subrange

Within the inertial subrange, the spectrum $E(k)$ of turbulent energy must be independent of viscosity and also independent of the processes that create the energy. In fact it can depend only on the wavenumber and the rate of dissipation ε. Dimensional analysis then shows

$$E(k) = A\varepsilon^{2/3}k^{-5/3} \ . \tag{9.19}$$

For one-dimensional spectra, this gives

$$S_u(k_1) = a\varepsilon^{2/3}k^{-5/3} \tag{9.20}$$

and similar spectra for v and w (the transverse spectra). Empirical data gives

$$a = 0.15 \quad \text{for} \quad S(k_1) \tag{9.21}$$

and

$$a = 0.20 \quad \text{for} \quad S(k_2) \quad \text{and} \quad S(k_3) \ . \tag{9.22}$$

Note that the ratio of transverse to longitudinal spectra is 4/3 for isotropic turbulence. When wavenumbers are defined as radians per unit length, these constants become 0.50 and 0.63.

Reynolds numbers are so large in the atmosphere that spectra must cover many decades of wavenumbers. Therefore it is useful to use a logarithmic scale for k_1. It is important to preserve the total area under the spectrum; if this is not done, the spectral density loses its meaning. We can do this by replacing $S(k_1)$ by $k_1 S(k_1)$.

Thus

$$k_1 S(k_1) \, d\ln k_1 \equiv S(k_1) \, dk_1 \,. \tag{9.23}$$

Such spectra are called logarithmic spectra. In the inertial subrange, (9.20) becomes

$$k_1 S(k_1) = a\varepsilon^{2/3} k_1^{-2/3} \,. \tag{9.24}$$

It needs to be remembered that in general, the maximum values of the spectrum and the logarithmic spectrum do not occur at the same wavenumber.

9.5 Surface Layer Velocity Component Spectra

The reciprocal of the wavenumber is the associated wavelength λ. For the largest eddies in the surface layer spectrum, this length is associated with the height above the ground. It is useful to introduce a dimensionless wavenumber n defined by

$$n = \frac{z}{\lambda} = k_1 z = \frac{f\,z}{\overline{u}} \,. \tag{9.25}$$

It should be noted that because of Taylor's hypothesis, n can be interpreted as the ratio of the height to the wavelength associated with the wavenumber.

The rate of dissipation in the surface layer is a rapidly varying function of height. It is useful to define a dimensionless dissipation by scaling ε with its neutral rate u_*^3/kz,

$$\varphi_\varepsilon \equiv \frac{kz\varepsilon}{u_*^3} \,. \tag{9.26}$$

Thus the dimensionless dissipation rate φ_ε is 1 for $z/L = 0$. Substitution of this into (9.23) gives the result that the scaled spectrum $G(n, z/L)$, defined by

$$G(n, z/L) \equiv \frac{nS(n)}{u_*^2 \varphi_\varepsilon^{2/3}} \,, \tag{9.27}$$

is independent of z/L for frequencies in the inertial subrange. Thus we see that for sufficiently high wavenumbers, the logarithmic spectra of the velocity components in the surface layer, when normalized as in (9.27), are functions only of n and are independent of z/L. This prediction is seen to be reflected in the observed spectra shown in Figs. 9.1 and 9.2. In the surface layer this independence, which characterizes the inertial subrange, prevails for values of n greater than about 5 under stable conditions and 1 or less under unstable conditions. These limits correspond to wavelengths between one-fifth the height and somewhat less than the height itself.

Small values of n represent the energy-containing eddy sizes. These are expected to be non-isotropic and show a dependence on the surface heating which

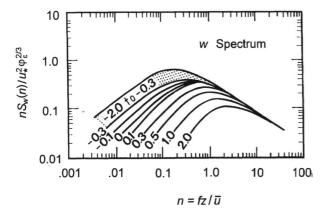

Fig. 9.1 Scaled surface-layer vertical velocity spectra for z/L values from +2.0 to −2.0 observed by Kaimal et al. (1972). Shading indicates the absence of any well defined dependence on z/L

produces their energy. Thus the normalized spectrum becomes more and more a function of z/L as the wavenumbers decrease. This prediction is seen to be borne out in the observed spectra for all three velocity components as long as z/L is greater than zero. For negative values of z/L the spectra of all three components are greater than the corresponding neutral spectra, but there appears to be no clear separation of observed spectra for different values of z/L. This fact suggests that during conditions of strong surface heating the spectra are influenced by factors other than distance from the ground. Presumably, this distance from the ground is replaced by the height of the mixed layer.

The spectra of the horizontal components display another interesting property. As one progresses from weakly stable to weakly lapse conditions, the spectra change in a discontinuous fashion. Not only do the individual spectra lose an orderly relationship to z/L, but also the region they occupy on the diagrams shown above is disconnected from the region occupied by the stable spectra at the higher wavenumbers. This behavior strongly suggests that at the lowest wavenumbers where most of the energy is contained, the winds are strongly influenced by the height of the mixed layer rather than distance from the ground, as we have noted before in connection with the lack of Monin–Obukhov similarity. Thus the excluded region on these diagrams and the lack of clear relation between the spectrum and z/L are indications of the shift from forced to free convection.

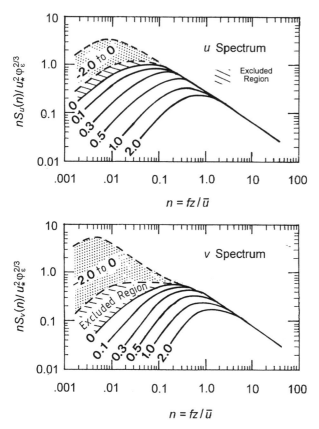

Fig. 9.2 Scaled spectra of the longitudinal (u) and (v) horizontal wind components in the surface layer, observed by Kaimal et al. (1972). Note the discontinuous progression of spectra in passing from stable to unstable conditions

9.6 Mixed Layer Velocity Component Spectra

Following the reasoning in the last section, it is expected that if the velocity component spectra are normalized by w_*^2 in the convectively mixed layer, they should be functions of z/h only. Within the inertial subrange the reasoning suggests that we redefine n as the nondimensional ratio of h divided by the wavelength or $k_1 h$. Equation (9.20) can then be written

$$nS_u(n) = a\varepsilon^{2/3}n^{-2/3}h^{2/3} \; . \tag{9.28}$$

We then normalize the equation by dividing both sides by

$$w_*^2 = \left(\frac{gH_0 h}{c_p\overline{\rho}\overline{T}_0}\right)^{2/3} \; . \tag{9.29}$$

Then

$$\frac{nS_u(n)}{w_*^2} = a\psi_\varepsilon^{2/3} n^{2/3} \tag{9.30}$$

where

$$\psi_\varepsilon = \frac{\varepsilon c_p \overline{\rho T}}{gH_0} \tag{9.31}$$

is a function of z/h as required by mixed layer similarity. Further, if we divide through by $\psi_\varepsilon^{2/3}$, the result should be a function only of n within the inertial subrange. Data analyzed by Kaimal et al. (1976), from a large experimental field program conducted in Minnesota, support this conclusion. (See Fig. 9.3.)

At lower wavenumbers the data support the conclusion that the normalized spectra of all three velocity components are functions of z/h.

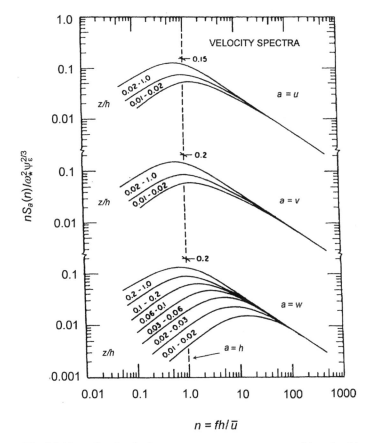

Fig. 9.3 Normalized velocity component spectra expressed in mixed-layer scaling coordinates. Wavelengths less than about $0.1h(n > 10h)$ lie within the inertial subrange. Reproduced from Kaimal et al. (1976)

9.7 Spectra of Scalar Quantities Including Temperature

Kaimal et al. have discussed the spectra of scalar quantities in a way that is quite analogous to those of the wind components. The non-dimensionalized spectrum of a quantity q can be expected to satisfy an expression

$$\frac{nS_q(n)}{q_*^2} = a_q \psi_q \left(\frac{z}{L}\right) n^{-2/3} , \tag{9.32}$$

where ψ_q is

$$\psi_q \left(\frac{z}{L}\right) = \varphi_q \varphi_\varepsilon^{-1/3} \tag{9.33}$$

and

$$\varphi_q \left(\frac{z}{L}\right) = \frac{kz}{q_*} \frac{\partial \overline{q}}{\partial z} . \tag{9.34}$$

One can then expect that the function

$$\Phi_q \left(n, \frac{z}{L}\right) \equiv \frac{nS_q(n)}{q_*^2 \varphi_q \varphi_\varepsilon^{-1/3}} \tag{9.35}$$

should be independent of z/L over the higher frequencies comprising the inertial subrange and vary with z/L at lower frequencies.

When q is the temperature, the dimensionless function φ_q is given by the Businger–Dyer formulas (5.22) and (5.23). It is generally believed that φ_q for other scalars can be calculated from the same formulas; while supporting data are few, there is almost no indication to the contrary. Wyngaard and Cote (1971) have suggested the formula

$$\varphi_\varepsilon = \varphi_m - \frac{z}{L} \cong 1 - \frac{z}{L} . \tag{9.36}$$

The normalized spectrum of temperature fluctuations in the surface layer is shown in Fig. 9.4. Observations of spectra of other scalar quantities are not available. However, it is likely that when properly normalized as described above, they would be identical to those shown in Fig. 9.4.

9.8 Cospectra and Quadrature Spectra

Wyngaard and Cote (1972) have used dimensional analysis to show that in the inertial subrange, the cospectrum must be proportional to k raised to the power $-7/2$. This contrasts with the spectra of energy and scalars that follow a $-5/3$

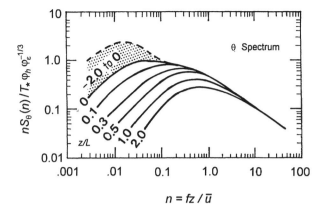

The figure plot shows axis labels: vertical axis $nS_\theta(n)/T_* \, \varphi_h \, \varphi_\varepsilon^{-1/3}$ with values 10, 1.0, 0.1, 0.01; and a label θ Spectrum. Horizontal axis values .001, .01, 0.1, 1.0, 10, 100. Curve labels z/L values: 0, 0.1, 0.3, 0.5, 1.0, 2.0, and -2.0 to 0.

$$n = fz / \bar{u}$$

Fig. 9.4 Scaled surface-layer temperature spectra after Kaimal et al. (1972)

power law. Thus cospectra decrease more rapidly with frequency in the inertial subrange than do the spectra.

This behavior of the cospectra is consistent with the expectation that as the wavenumbers increase, the turbulence tends to become isotropic. If turbulence were completely isotropic, the covariances would all be zero.

In general the approach to the scaling of cospectra follows the same procedure as that of spectra, and will not be discussed here. The normalized cospectra for uw and $w\theta$ are shown in Fig. 9.5. Comparison of these with the spectra for u, w, and θ show that not only do the cospectra fall off more rapidly at high wavenumbers but also the wavenumber of the maximum shifts towards lower values than the maxima of either of the spectra. Thus the fluxes are carried on quite selectively by the largest eddies in the spectrum. This fact is of considerable benefit for the task of measuring fluxes by the *eddy-correlation* method. In this method, rapidly responding instruments are used to measure the turbulent fluctuations and to evaluate their mean products. This method requires instrumentation that is able to respond faithfully to the fluctuations that are responsible for the covariance, and the higher the required frequency of response, the higher becomes the cost of providing it. Fortunately, it is relatively inexpensive to measure the important frequencies because these are the smallest. Provided one is able to measure the important frequencies of the cospectrum, and know the cutoff frequency of the instrumentation, it is possible to use the non-dimensional cospectra to correct for the portions of the covariance that cannot be measured directly.

Almost nothing is known about quadrature spectra. Their interpretation has been discussed by Panofsky and Dutton (1984) in connection with the variables u and w. At any particular frequency, the sign of the quadrature spectrum indicates which of the two variables leads the other in their out-of-phase relationship. At the bottom of an eddy, positive w precedes positive u while near the top the reverse is true. Lumley and Panofsky (1964) showed from observations made at Brookhaven Laboratory that in stable conditions, the lowest level was at the

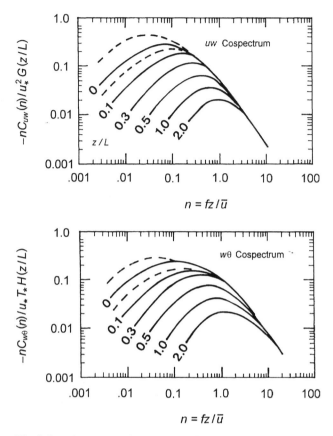

Fig. 9.5 Scaled spectra for uw and $w\theta$ observed by Kaimal et al. (1972)

bottom of the passing eddies, and the 91 m level was near the top. In unstable conditions both levels were near the bottom.

9.9 Problems

1. Under neutral conditions, the maximum value of the logarithmic spectrum of vertical velocity occurs at a value of n equal to about 0.5. How does the wavelength compare to the height at this value of n?

2. For the situation described in # 1 above, does the maximum value of the ordinary spectrum $S_w(n)$ occur at a smaller or larger value of n?

3. Why is it justified to integrate (9.3) over only half the range of (9.1)?

4. Explain why the cospectrum of two different velocity components must be zero when the turbulence is isotropic.

5. Give an example of a commonly seen visual phenomenon that supports Taylor's hypothesis.

6. How does the Kolmogorov microscale vary with height under neutral conditions? If the stratification changes to lapse with no change of u_* or height, will the microscale get larger or smaller?

7. Find the ratio of $nS(n)$ for water vapor to that of temperature at the same height. Assume that the constant in (9.32) is the same for water vapor and heat.

8. Demonstrate using dimensional analysis that the energy spectrum in the inertial subrange is proportional to wavenumber to the minus five-thirds power.

9. The dissipation spectrum is small enough in the inertial subrange that it can be neglected. However, it is not zero. If you were to plot the logarithmic dissipation spectrum on the graph shown in Fig. 9.1, what would be its slope?

10 Turbulent Diffusion from Discrete Sources

The forms that smoke plumes exhibit are a result of the action of atmospheric turbulence, and much can be learned about the nature of turbulence by observing them. Conversely, knowledge about turbulence, acquired over a long period of measurement and consideration, is now being used successfully to help predict the features of smoke plume movements. The practical result of such applications is a growing ability to predict in advance how the byproducts of existing or projected industrial enterprises affect the air quality of surrounding areas.

10.1 Morphology of Smoke Plumes

The shapes displayed by smoke plumes are related to the lapse rate and to the wind speed. With high winds and/or near-neutral lapse rates. the turbulence energy is mainly generated by mechanical production, and the eddies are then small. With moderate wind speeds, such eddies cause the wind vane to oscillate rapidly back and forth through a small angle. When the lapse rates are large and the wind speeds are light, convection dominates the energy generation process more and more and larger and larger eddies are added to the spectrum. If the wind speeds are light, such eddies produce large variations of vertical and horizontal wind direction with a duration of 20 minutes or more.

The smallest eddies cause a smoke plume to expand and become less distinct. The larger eddies, those whose size is larger than the width of the plume, tend to move the entire plume from side to side, as well as up and down, causing it sometimes to take on an irregular snakelike appearance. We describe such plumes as *looping*. Conditions of this kind occur best with weak winds, sunny skies, and surface conditions that tend to minimize the rate of evaporation. Plumes that are affected mainly by small eddies lack the snakelike characteristics and simply spread slowly as they drift downwind. We call such behavior *coning* because of the conical appearance of the plume. These plumes occur with high wind speeds or with lower wind speeds when the sky is overcast and the surface heating weak.

A third category of smoke plume behavior, called *fanning*, occurs on clear nights with weak winds. Under these conditions eddies tend to become horizontal; there is virtually no vertical spreading of the plume. Such plumes slowly fan out in a horizontal plane as they move downwind from the source. The best time to see such plumes is in the early morning around the time of sunrise.

Fanning plumes are composed of heavy concentrations of pollutant but are not usually dangerous at night because of their high elevation above the ground. Danger increases after sunrise as convectively driven turbulence begins to rise from the surface. When the convective layer reaches the plume level, the large concentrations may be transported abruptly to the ground. Such a condition is called *fumigation*. Some of the worst pollution episodes have occurred in this type of situation.

It is easy to get the impression that the motions that produce big loops are highly chaotic. Actually the individual particles of the plume tend to move in nearly straight lines, especially if they are not close to the ground. This fact can be observed by concentrating attention on one of the denser blobs leaving a chimney and following it as it drifts downwind. The loopy appearance develops from the more rapid changes of directions of the paths of successive particles as they get caught up in the procession of different eddies that move past the chimney.

When a loop reaches the ground, a high concentration of smoke is likely to be experienced by inhabitants at the impacted site. Since loops are associated with the largest eddies and light wind conditions, these high concentrations may persist for several minutes at a time. For the unfortunate people who are thus exposed, it is a small consolation to know that there are similarly long periods when the loops carry the smoke upward or to other locations at the surface.

Clean-air legislation has mainly been concerned with the average concentration over a one-hour or longer period. The greatest concern is the maximum average concentration that occurs at the ground. If the chimney is high and the flow past it is not greatly influenced by local obstructions, the average ground concentration nearby is quite small. It takes time for the plume to be transported downward and by then the wind has carried it further away. Likewise, very far away, maybe 10 or 20 km, the concentrations are also very small because of the large amount of vertical and horizontal dispersion that has gone on over the long distance of travel. Somewhere in between there is usually a point where the ground concentration is a maximum.

With respect to short-term, ground-level pollution concentrations near their source, high chimneys are advantageous because they shift the point of maximum concentration further downwind allowing more time for the concentrations to be reduced by turbulent dispersion. Many newer power plants have chimneys exceeding 1000 feet (300 m) in height. The effective height of these chimneys is actually considerably greater because the heat and momentum of the effluent gases carry them upward for some distance before the center of the plume levels off.

Large effective emission heights are not an unmitigated blessing. The pollutants that are prevented from reaching the surface immediately downwind of the source become part of the widespread burden of resident atmospheric pollutants. Before they are eventually deposited on the surface they may undergo transformation into a multitude of undesirable chemical byproducts.

10.2 Continuity Principles

Even without any knowledge or understanding of the turbulence that is responsible for the shape and spreading of smoke plumes, it is possible to derive quite a bit of information about them. What we shall explore is the consequence of the principle of conservation of matter. In so doing, we assume that the chemical transformation of the effluent is negligible during the rather short lifetime of the plume.

We consider a particle which leaves its source at some instant of time and moves in response to the turbulent motions of the air. We take the mean downwind direction to be the x-direction, which is also often called the longitudinal direction. Perpendicular to this are the horizontal lateral direction y and the vertical direction z. We define the probability density $F(x)$ such that the probability that the particle will lie at a value of x between x and $x + dx$ is $F(x)dx$. If the particle does not disappear or change to something else, then

$$\int_{-\infty}^{+\infty} F(x)\, dx = 1 . \tag{10.1}$$

We can define similar probability densities $G(y)$ and $H(z)$. An important assumption, usually believed to be justified, is that the probability densities in the different directions are independent of each other; that is, the probability that the particle lies in a certain range of x values is independent of its position in the y and z directions. In this case, the probability that a particle lies simultaneously between x and $x + dx$, y and $y + dy$, and between z and $z + dz$ is the product of the separate probabilities. Again, if the particle's existence is assured, we have at any subsequent time after its release

$$\int_{-\infty}^{+\infty} \int_{-\infty}^{+\infty} \int_{-\infty}^{+\infty} F(x)\ G(y)\ H(z)\, dx\, dy\, dz\ =\ 1 . \tag{10.2}$$

We consider an instantaneous point source of strength Q. This could be a certain number of particles, a certain mass of some species of pollutant, or a volume of some gas, for example. We define the concentration χ as the probability density of the pollutant in the same units, that is, number of particles per unit volume, mass per unit volume, or volume of pollutant per unit volume of space. Thus the expected amount of pollutant in a volume of dimensions dx, dy, dz located at x, y, and z is $\chi dx\, dy\, dz$. From these definitions we have

$$\chi(x, y, z) = Q\ F(x)\, G(y)\, H(z) \tag{10.3}$$

and

$$Q = \int_{-\infty}^{+\infty} \int_{-\infty}^{+\infty} \int_{-\infty}^{+\infty} \chi(x, y, z)\ dx\, dy\, dz . \tag{10.4}$$

There is much more to be said about the form of these functions later on. For now it is important to recognize that these equations must be satisfied without regard to the form of the function. There is one special case that we shall find useful. If a cloud is confined between two plates, as for example the ground underneath and an inversion above, then a long time after the release from a point source between the two plates, the concentration will approach a limit in which the concentration is; uniformly distributed as a function of height between the plates; zero above and below the plates. If the separation of the two plates is D, then it is easily shown that $H = 1/D$. (See Problem 1 at the end of the chapter.)

It should be noted that when the wind changes with height the independence of the probability distributions cannot be defended. It has also been assumed that the substance being diffused or transported is conserved. There are many processes in the atmosphere that undermine this assumption, especially when long time periods are concerned: sticking of particles to the ground surface; rain-out or wash-out of particles by precipitation; and chemical transformation of one species into another are some common examples.

Most instantaneous sources that have to be dealt with are effectively single points, or can be dealt with as a superposition of several independent point sources. An example would be an accidental release of material at a plant or a ruptured vehicle. While such sources are finite in size, they are effectively points in comparison with the volume of the subsequent cloud that must be dealt with. If the size of the initial cloud is large enough to require attention, it can sometimes be dealt with by assuming that it represents the evolution from a virtual point source at an earlier time.

A second class of problems is the continuous point source – a smoke plume issuing from a chimney, for example. While the diameter of the source may have important consequences on the buoyant rise of the plume, it is usually quite negligible compared with the width and distances of the areas that are affected. Again, the concentration of pollutant in the resultant cloud can be expressed in terms of a joint probability density.

We can apply the results of the instantaneous point source problem by superimposing a sequence of such sources spaced at intervals of dt. It is necessary to redefine the source strength Q as the emission rate per unit time, so that the discrete point source strength becomes Qdt. Usually the simplifying assumption is made that there is no dispersion in the direction of the mean wind, so that the emission of each source is confined within a slab of width udt at a value of x equal to u times the time since emission, as shown schematically in Fig. 10.1. Using the special case referred to earlier of the probability density of a uniformly distributed pollutant between two plates, we find $F(x)$ for the concentration within the slab is

$$F(x) = \frac{1}{\overline{u}\, dt} \; . \tag{10.5}$$

We can now apply (10.3) to get the total concentration, noting that the infinitesimal dt occurs in both the numerator and denominator and thus is no problem as $dt \to 0$.

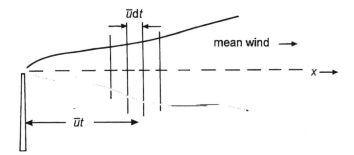

Fig. 10.1 Approximating a plume by a succession of point sources

$$\chi(x, y, z) = \frac{Q}{\overline{u}} G(y) H(z) \ . \tag{10.6}$$

Even without further information on the form of $G(y)$ and $H(z)$, it can be said of all smoke plumes that the concentration of pollutant everywhere is proportional to the emission rate and inversely proportional to the wind speed.

Another type of problem occasionally encountered is a continuous line source. An example of this type of problem is traffic along a highway which, on average, exhibits a uniform output of a pollutant per unit length and time. With little loss of generality we can assume that the line is perpendicular to the mean wind, that is, in the y-direction. Again we must redefine source strength as the amount of the pollutant per unit time and per unit length along the line. For a small length dy of the line then, the amount of pollutant released per unit time is Qdy. Since the line is assumed to be infinite in length, the amount of this release that is diffused away laterally is matched by the amounts received from all the other incremental releases along the line. Thus it is equivalent to completely disregard lateral diffusion and assume that all the material released in the segment dy stays within this segment as it is moves downwind and upward. Thus for each segment, $G(y)$ is equal to $1/dy$. Accordingly we have

$$\chi(x, z) = \frac{Q}{\overline{u}} H(z) \ , \tag{10.7}$$

and again we find that the concentration downwind of the line is proportional to the amount of material released per unit length of the line and inversely proportional to the mean wind speed.

10.3 Fickian Diffusion

Except for one or two trivial cases, we have avoided the question of the form of the probability density functions and the parameters that determine them. Clearly this is the most important problem in predicting pollutant concentration, since when these functions are known as functions of time and space, it is possible to estimate the mean concentration of pollutant at any location and at any time.

The earliest approach to the solution of this problem was to apply K-theory. We have already seen that the rate of change of the concentration of a property depends on the divergence of the three-dimensional flux of the property, and that, under appropriate restrictions, this flux is proportional to the gradient of the mean concentration. In a three-dimensional (i.e. inhomogeneous) environment, the equation governing changes in the mean concentration is

$$\frac{\partial \chi}{\partial t} + \overline{u}\frac{\partial \chi}{\partial x} = \frac{\partial}{\partial x}\left(K_x \frac{\partial \chi}{\partial x}\right) + \frac{\partial}{\partial y}\left(K_y \frac{\partial \chi}{\partial y}\right) + \frac{\partial}{\partial z}\left(K_z \frac{\partial \chi}{\partial z}\right) . \tag{10.8}$$

This equation assumes that the mean wind is in the x-direction. K_x, K_y, and K_z are the exchange coefficients in the x-, y-, and z-directions, respectively. Actually this equation is an oversimplification of the most general possible relationship, as was originally shown by Ertel, since it assumes that the flux in each direction is independent of the flux in the other directions. Ertel showed that in the general case the scalar quantities K_x, K_y, and K_z must be replaced by a symmetric tensor of rank 4. Although we do not try to be completely general, we should avoid thinking of K_x, K_y, and K_z as components of a vector, and we have been careful not to use the Einstein summation convection in dealing with these quantities.

Before this equation can be solved it is necessary to specify the distribution of the three exchange coefficients in x, y, z, and time. The simplest case, and the one that is most easily solved is to assume that all the coefficients are constant and equal. If we put the constant exchange coefficient equal to K, the right side of (10.8) is K times the Laplacian of χ. If we allow each K to be different but constant we have the Fick equation

$$\frac{\partial \chi}{\partial t} + \overline{u}\frac{\partial \chi}{\partial x} = K_x \frac{\partial^2 \chi}{\partial x^2} + K_y \frac{\partial^2 \chi}{\partial y^2} + K_z \frac{\partial^2 \chi}{\partial z^2} . \tag{10.9}$$

If the mean wind is constant one can eliminate the second term of (10.8) by allowing the coordinate system to move with the mean wind. With this assumption we can arrive at solutions for a variety of problems. Alternatively, one can consider a steady state solution for a continuous source in a mean wind $\overline{u}$, as was done by Roberts (1923).

We consider first an instantaneous point source and move the coordinate system with the mean wind. The equation to be solved is

$$\frac{\partial \chi}{\partial t} = K_x \frac{\partial^2 \chi}{\partial x^2} + K_y \frac{\partial^2 \chi}{\partial y^2} + K_z \frac{\partial^2 \chi}{\partial z^2} \tag{10.10}$$

subject to the conditions

$$\chi = 0 \quad \text{at} \quad t = 0 \quad x, y, z \neq 0, \, 0, \, 0$$
$$\chi \to 0 \quad \text{as} \quad t \to \infty \quad \text{for all } x, y, z$$

and

$$\int_{-\infty}^{+\infty} \int_{-\infty}^{+\infty} \int_{-\infty}^{+\infty} \chi(x, y, z)\mathrm{d}x\mathrm{d}y\mathrm{d}z = Q$$

where Q is the total mass released at time $t = 0$.

The solution, which can easily be verified by substitution into the differential equation, is

$$\chi = \frac{Q}{8\,(\pi t)^{3/2}\,\left(K_x K_y K_z\right)^{1/2}}$$

$$\times \exp\left(\frac{-x^2}{4K_x t}\right) \exp\left(\frac{-y^2}{4K_y t}\right) \exp\left(\frac{-z^2}{4K_z t}\right) . \tag{10.11}$$

This can be put in the form of an error function by setting

$$\sigma_x^2 = 2K_x t ; \quad \sigma_y^2 = 2K_y t ; \quad \sigma_z^2 = 2K_z t . \tag{10.12}$$

Then

$$\chi = \frac{Q}{(2\pi)^{3/2}\,\sigma_x \sigma_y \sigma_z} \exp\left(\frac{-x^2}{2\sigma_x^2}\right) \exp\left(\frac{-y^2}{2\sigma_y^2}\right) \exp\left(\frac{-z^2}{2\sigma_z^2}\right)$$

$$= Q\,F(x)\,G(y)\,H(z) \tag{10.13}$$

where

$$F(x) \equiv \frac{1}{\sqrt{2\pi}\sigma_x} \exp\left(\frac{-x^2}{2\sigma_x^2}\right) ; G(y) \equiv \frac{1}{\sqrt{2\pi}\sigma_y} \exp\left(\frac{-y^2}{2\sigma_y^2}\right) ;$$

$$H(z) \equiv \frac{1}{\sqrt{2\pi}\sigma_z} \exp\left(\frac{-z^2}{2\sigma_z^2}\right) . \tag{10.14}$$

It can quite easily be shown that the integral of each of these over all values of the argument is unity as required by continuity. We have seen how the results of a solution of the instantaneous source can be broadened to a continuous point source. The result is

$$\chi = \frac{Q}{\overline{u}} G(y)\,H(z) = \frac{Q}{2\pi\overline{u}\sigma_y \sigma_z} \exp\left[-\left(\frac{y^2}{2\sigma_y^2} + \frac{z^2}{2\sigma_z^2}\right)\right] . \tag{10.15}$$

The quantities indicated by σ_z are a measure of the width of the cloud or plume. The visible diameter is usually about four times the value of σ_z. An important aspect of the solution is how the values of σ_z increase with time or equivalent distance downwind of the source. Observations show σ_y and σ_z increase in proportion to t^n where $0.75 < n < 1.0$. This observed behavior is quite different from the predictions of the Fick equation, which indicate $n = 0.5$.

There are many reasons why K-theory fails to describe properly the diffusion in plumes of this kind. In this case, the failures appear to be associated with the fact that K-theory is based on the assumption that the diffusion is carried on by the energy-containing eddies, that is, by the longest wavelength eddies in the spectrum. However, when the cloud is small, the entire spectrum of the eddies causes the plume to expand. Eddies smaller than the cloud make the cloud itself expand. More importantly, however; eddies that are larger than the cloud move the entire cloud from side to side, and when the various instantaneous clouds are averaged, this motion contributes to the average plume width. Thus initially the plume expands more rapidly than it does when it gets larger, and can eventually only be moved by the largest eddies of the spectrum.

10.4 The Gaussian Distribution Function

The form of the function that the Fick equation predicts for the probability densities $F(x)$, $G(y)$, and $H(z)$ in (10.14) is known as the *Gaussian* distribution. This form of the shape of plumes and clouds is consistent with most experimental data if sufficient allowance is given to sampling irregularities, which must be expected for records of finite duration. This function also is a good approximation to the probability density of the velocity components that were discussed in Chap. 8.

To discuss the properties, let us consider the concentration of material in a plume along a horizontal line perpendicular to the direction of the wind, situated some distance downwind of a continuous source. The relative concentration of pollutant along this line is then described by the probability density function

$$G(y) = \frac{1}{\sqrt{2\pi}\sigma_y} \exp\left(\frac{-y^2}{2\sigma_y^2}\right) . \tag{10.16}$$

The form of this distribution is shown in Fig. 10.2. We let y represent the distance from the centerline where the probability density is a maximum, and we shall measure this distance in units of σ_y. About 2/3 of all the pollutant situated along this transverse line lies within a distance of one standard deviation σ_y of the center line, and a little more than 95% is located within a distance of $2\sigma_y$ from the centerline. Two standard deviations is a practical measure of the plume half-width.

Since the form of the Gaussian distribution seems to describe quite well the form of the observed concentration, and is consistent with some theoretical considerations, it is commonly used as the basis for predicting concentrations at any particular place. Before this can be done, however, it is necessary to have a method of predicting the standard deviations σ_y and σ_z. As we have seen at the beginning of this chapter that these parameters are sensitive to wind speed and surface heating rate. There are two approaches to the estimation of σ_y and σ_z. The first is to examine the Lagrangian statistics of particle motions. The second is to use the results of experiments to estimate σ_y and σ_z. We shall discuss these in the order listed.

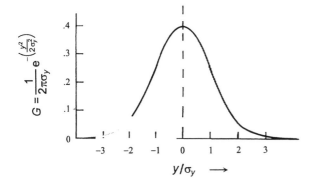

Fig. 10.2 The Gaussian distribution function often used as a model for relative cross-wind concentration

10.5 Taylor's Diffusion Equation

An important contribution to our understanding of turbulent diffusion was made by Taylor (1935). The theory recognized diffusion as a continuous process, in contrast to exchange theory, which is based on a model of discrete mixing events interspersed with undisturbed displacements over paths of undetermined lengths. While Taylor's equation falls short of providing a complete theory of diffusion, it focuses attention on those processes that are most important and on the statistics that must be known for further progress.

We shall discuss only dispersion in the y-direction, but the outcome can also be applied to the z-direction. The turbulence is presumed to be stationary. Particles are released from a source one by one, and each particle's motion is independent of that of its predecessor. The aim is to predict the variance of the y values of all the particles at a time interval T after leaving the source.

We define the Lagrangian autocorrelation $R(\xi)$ by the equation

$$R(\xi) \equiv \frac{\overline{v(t)\,v(t+\xi)}}{\overline{v^2}} \tag{10.17}$$

where $v(t)$ is dy/dt for a single particle at some time t, and $v(t+\xi)$ is the value for the same particle ξ seconds later. The average is extended over all the particles released at the same point. Under *stationary* conditions, the Lagrangian autocorrelation is a function of ξ only. Its value for $\xi = 0$ is unity.

We can integrate $R(\xi)$ from the moment of release ($t = 0$) to a time t seconds later:

$$\overline{v^2} \int_0^t R(\xi)\,\mathrm{d}\xi = \overline{v(t)\,y(t)} = \frac{1}{2}\frac{\mathrm{d}}{\mathrm{d}t}\,\overline{y^2}(t)\ . \tag{10.18}$$

A second integration, treating t as a dummy variable over the range 0 to T, gives, in principle, the mean squared value of y.

$$\overline{y^2}(T) = \int_0^T \frac{d}{dt}\, \overline{y^2}dt = 2\sigma_v^2 \int_0^T dt \int_0^t R(\xi)\,d\xi \ . \tag{10.19}$$

In this equation we have replaced, $\overline{v^2}$ by our previously used symbol σ_v^2. T now represents the time interval after leaving the point source. The statistic $\overline{y^2}$ is not exactly what we have previously defined as σ_y^2: the latter refers to the standard deviation of the values of y of all the particles that lie a distance $X = \overline{u}T$ downwind of the source, while $\overline{y^2}$ refers to all the particles that have travelled a time T since leaving the source. Some of these will be situated closer and some further from the source than distance X. The difference is seldom very great, and for most purposes we can identify the two statistics as the same. Thus we may interpret Taylor's equation as

$$\sigma_y^2(X) \cong 2\sigma_v^2 \int_0^T dt \int_0^\xi R(\xi)\,d\xi \tag{10.20}$$

where X is the mean distance from the point of release of all the particles that have traveled an interval of time T since leaving the source.

The problem of diffusion is thus reduced to finding R as a function of ξ. There are two special cases of interest. The first concerns the dispersion rate immediately following the time of release from the source. We may define the word *immediately* as T small enough that we can put $R(\xi) \approx 1$. In this case, (10.20) gives

$$\overline{y^2} = \sigma_v^2 T^2 \quad \text{or} \quad \sigma_y \approx \sigma_v T = \frac{\sigma_v X}{\overline{u}}\ . \tag{10.21}$$

Thus initially σ_y is a linear function of time T or X. Initially the plume spreads linearly with a constant angular width, which is proportional to σ_α, the standard deviation of wind directions measured by a wind vane situated at the point of release. If the angles are small, this standard deviation is given quite well by

$$\sigma_\alpha \cong \frac{\sigma_v}{\overline{u}} \quad \text{and} \quad \sigma_y \cong \sigma_\alpha X\ . \tag{10.22}$$

The second special case concerns the behavior a very large distance away from the source. We shall assume that an integral time scale, in this case the Lagrangian time scale defined by

$$T_L \equiv \lim_{t \to \infty} \int_0^t R(\xi)\,d\xi \tag{10.23}$$

exists. At a sufficiently long time after release, the Taylor equation becomes

$$\sigma_y^2 = 2\sigma_v^2 T_L T \tag{10.24}$$

or

$$\sigma_y \propto \sqrt{T}\ . \tag{10.25}$$

This limiting behavior is identical to the result obtained from the Fick equation. The reasons for this behavior will be explored in the next section.

10.6 Spectral Representation of Taylor's Equation

The Lagrangian autocorrelation can be used to obtain the spectrum $F_v(n)$ of the variance $\overline{v^2}$ as function of n, the number of cycles per unit time experienced by a particle as it moves along its trajectory. The spectrum is the Fourier transform of the autocorrelation

$$F_v(n) = 4\overline{v^2} \int_0^\infty R_v(t) \cos(2\pi nt) \, dt \, . \tag{10.26}$$

The cosine transform has been used because $R(t)$ is an even function of t. The inverse transform gives

$$\overline{v^2} R_v(t) = \int_0^\infty F(n) \cos(2\pi nt) \, dn \, . \tag{10.27}$$

The left side of the last equation appears in Taylor's diffusion equation, and it is useful to replace it by the function on the right side. To this end it is expedient to use an alternative form of Taylor's equation that was derived by Kampé de Férier (1939):

$$\overline{y^2} = 2\overline{v^2} \int_0^T (T - t) R_v(t) \, dt \, . \tag{10.28}$$

This has the advantage that one of the integrations has been performed explicitly. Substitution then gives

$$\overline{y^2} = 2 \int_0^\infty dn \int_0^T (T - t) F_v(n) \cos(2\pi nt) \, dt \, . \tag{10.29}$$

This equation is then integrated with respect to t, giving an equation of the form

$$\overline{y^2}(T) = 2 \int_0^\infty F_v(n) f(n, T) \, dn \tag{10.30}$$

in which it can be shown with some work that

$$f(n, t) = \frac{T^2}{2} \frac{(\sin \pi nT)^2}{(\pi nT)^2} \, , \tag{10.31}$$

so that we have finally

$$\overline{y^2} = T^2 \int_0^\infty F_v(n) \frac{(\sin \pi nT)^2}{(\pi nT)^2} \, dn \, . \tag{10.32}$$

Before attempting to interpret this result, let us take a special case in which T is very small. This corresponds to the initial spreading of the plume just after leaving the source. In this case, the weighting function

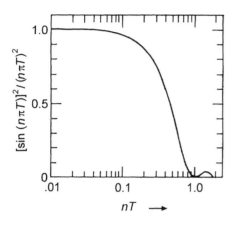

Fig. 10.3 Weighting function for the proportion of the energy spectrum responsible for lateral diffusion as a function of time T after release

$$\frac{(\sin \pi n T\,)^2}{(\pi n T\,)^2} \tag{10.33}$$

shown in Fig. 10.3 is unity, and the integral is $\overline{v^2}$, the total variance of the y-components of the turbulent velocities. Thus we see that initially the entire energy spectrum participates in the spreading of the plume.

The function defined in (10.33) and shown in Fig. 10.3 is a filter that defines the portion of the Lagrangian spectrum that is responsible for the dispersion at time T after release. Fig. 10.3 shows that only those frequencies for which the product nT is small are effective. Thus n must be small compared to $1/T$, and so as T increases, the eddy sizes that are effective in increasing the width are more and more restricted to the longest periods.

10.7 Stability Parameters

We have seen that the single most important parameter in determining the shape of the plume extending downwind from a continuous point source is the standard deviation σ_α of the horizontal wind directions. This statistic is usually well approximated from the ratio $\sigma_v/\overline{u}$. In Chap. 8 it was shown that is a sensitive function of stability, and is independent of height. The task is to translate our knowledge of the variation of σ_v with stability and wind speed into a meaningful parameter that will enable one to estimate σ_α from easily measured or estimated meteorological input data.

The standard deviation of the wind direction angles is relatively easy to measure from a wind vane installed at the point of release. However, it is not a part of the normally recorded weather observation, and usually has to be estimated from other data. If we combine our knowledge of mean wind and σ_v as functions of height and stability, we have

$$\sigma_\alpha = \frac{\sigma_v}{\bar{u}} = \frac{k(12 - 0.5h/L)^{1/3}}{\ln(z/z_0) - \psi(z/L)} \tag{10.34}$$

after using (5.17) and (8.22). In this equation h is the mixed layer height and k is the von Karman constant.

As long as L, z, and h remain constant, σ_α is independent of the wind speed. Under neutral conditions σ_α depends only on height. Since σ_v is independent of height and $\bar{u}$ increases with height, σ_α tends to decrease upward. At moderate heights and surface roughness values, σ_α is typically about $10°$.

With a given rate of surface heating, L is strongly dependent on u_*. Under unstable conditions L becomes negative and with light winds its magnitude becomes small. The result is to increase the numerator of (10.34) and decrease the denominator. In such cases, increasing the wind speed is seen to change σ_α back toward the neutral value. Under stable conditions and light winds, σ_α becomes small and may become zero if turbulence ceases completely. (Often under such conditions there is a certain amount of nonturbulent meandering.) With stronger winds in stable conditions, σ_α increases toward the neutral value. We see then that a meaningful parameter for estimating the initial rate of spreading of a plume must incorporate both the heating rate and the wind speed.

Equation (10.34) and a similar one for the standard deviation for the vertical angle fluctuations could be useful for predicting the initial dispersion of a plume, in the absence of direct observations. However, we seldom have precise knowledge of the heating rate, and can only make rough estimates of u_* and h. For this reason, it is helpful to use a classification scheme devised by F. Pasquill which, though less precise, incorporates the essence of the quantitative equation (10.34). The intent is to identify the meteorological conditions that lead to the ranges of σ_α defined in Table 10.1.

Table 10.1 Standard deviations of wind angles for various stability classes

Pasquill stability class	σ_α
A Extremely unstable	25
B Moderately unstable	20
C Slightly unstable	15
D Neutral	10
E Slightly stable	5
F Moderately Stable	2.5

Pasquill has shown that these relationships can generally be achieved by the use of Table 10.2. In this table there is no class that corresponds to very light wind speeds on clear nights. This condition generally results in a total lack of vertical turbulent motions and might reasonably be assigned a letter G. Under such conditions one expects to find a fanning plume.

One of the shortcomings of the Pasquill classification is that it does not take account of the mixed-layer height or roughness, both of which affect σ_α as seen

Table 10.2 Pasquill stability types

A	Extremely unstable	D	Neutral
B	Moderately unstable	E	Slightly stable
C	Slightly unstable	F	Moderately stable

Surface wind speed $\mathrm{m\,s^{-1}}$	Daytime insolation			Night-time Conditions	
	Strong	Moderate	Slight	Thin overcast or cloudiness $\geq 4/8$	Cloudiness $\geq 3/8$
< 2	A	A–B	B		
2	A–B	B	C	E	F
4	B	B–C	C	D	E
6	C	C–D	D	D	D
> 6	C	D	D	D	D

in (10.34). Increasing the roughness or decreasing the height both have the same effect as increasing the instability. Many users of the classification compensate for rough locations by moving to the next more unstable class (e.g. from C to B). There does not appear to be any effort on the part of users to include the height of the mixed layer in determining the stability class, probably because it is seldom known precisely.

10.8 Gaussian Plume Models

Both theory and observation suggest that the probability density functions $G(y)$ and $H(z)$ conform acceptably well to the Gaussian form. For a continuous source of strength Q units per unit time, the concentration at a point x, y, z is given by the equation

$$\chi(x,y,z) = \frac{Q}{2\pi\bar{u}\sigma_y\sigma_z} \exp\left(\frac{-y^2}{2\sigma_y^2}\right) \exp\left(\frac{-z^2}{2\sigma_z^2}\right) . \tag{10.35}$$

While x does not appear explicitly in this equation, it must be recognized that σ_y and σ_z are functions of x that must be determined before the concentration can be calculated. In writing (10.35) it is assumed that the source is situated at $z = 0$, so that z represents the vertical distance above or below the height of the source. Usually we prefer to define z as the height above the ground, in which case, if h_e is the effective height of the source, (10.35) becomes

$$\chi = \frac{Q}{2\pi\bar{u}\sigma_y\sigma_z} \exp\left(\frac{-y^2}{2\sigma_y^2}\right) \exp\left(\frac{-(h_e - z)^2}{2\sigma_z^2}\right) . \tag{10.36}$$

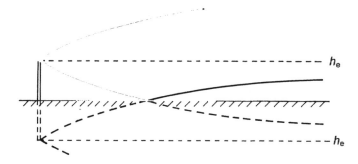

Fig. 10.4 Reflected plume created by use of a virtual source at $z = -h_e$

Clearly this equation should not be used for $z < 0$, but it is not clear how to deal with those particles that impact the ground. If we could assume that all the particles that touch the ground are absorbed, we just ignore the portion of the plume where $z < 0$. However, if we accept the opposite extreme, and assume that all the particles are reflected at the surface, then the concentration at heights above the surface will be higher than those given by (10.36). This case is usually dealt with by simply reflecting the subsurface plume into the upper space and superimposing it on the unreflected plume. The latter case is equivalent to adding a second virtual source of equal strength at the point $0, 0, -h_e$ as illustrated in Fig. 10.4. The concentration at any point above or at the surface is then given by

$$\chi = \frac{Q}{2\pi \overline{u} \sigma_y \sigma_z} \exp\left(\frac{-y^2}{2\sigma_y^2}\right)$$

$$\times \left[\exp\left(\frac{-(h_e - z)^2}{2\sigma_z^2}\right) + \exp\left(\frac{-(h_e + z)^2}{2\sigma_z^2}\right)\right] . \tag{10.37}$$

Probably the true concentration lies somewhere between those given by (10.36) and (10.37). Since the users of the model must generally defend their results to regulatory agencies, they usually prefer to use (10.37) because it is more conservative. For the case of concentration at the ground, the most frequent application, (10.37) gives twice the value of (10.36).

Due to the finite exit velocity of the effluent from the chimney, and its relatively high temperature, plumes rise buoyantly for a certain distance before levelling off. The amount of rise depends on the amount of buoyancy and on meteorological parameters, especially the wind speed and the stability: the amount of rise is greatest at low wind speed. Because of this plume rise, the effective height of the plume, which must be used in the Gaussian plume formulas above, is larger – often considerably larger – than the actual height of the chimney. A general-purpose formula for estimating the amount of plume rise Δh to expect at a distance x within about 10 stack heights downwind of the stack (Briggs (1969)) is

$$\Delta h = 1.6 F^{\frac{1}{3}} \overline{u}^{-1} x^{\frac{2}{3}} , \tag{10.38}$$

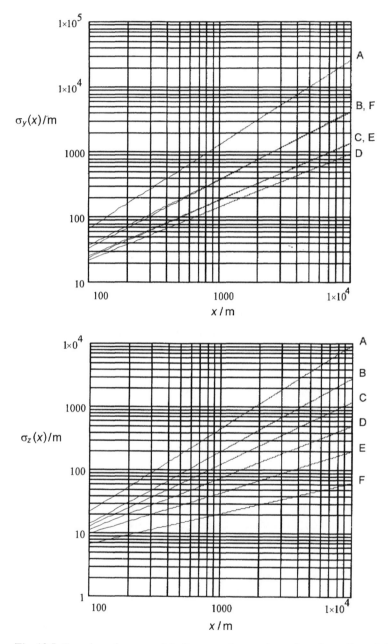

Fig. 10.5 Gaussian plume model diagrams for σ_y (*upper*) and σ_z (*lower*) for effective release heights of 100 m over rough terrain based on compilations by Geiss et al. (1981). Letters refer to the Pasquill stability class

in which $\overline{u}$ is the mean wind speed at the top of the stack, and F is the buoyancy flux parameter (dimensions L^4T^{-3}) defined by

$$F \equiv \frac{gQ_H}{\pi c_p \rho \overline{T}}$$ (10.39)

in which Q_H is the amount of heat emitted from the stack per unit time. A comprehensive theory of plume rise applicable to more specific conditions has been developed by Briggs (1975).

The most widely used method of determining σ_y and σ_z is based on a set of diagrams devised by Pasquill and Gifford (see Gifford (1968)). The original diagrams were based on measurements of tracers released at the surface over smooth terrain. These diagrams are still in widespread use, generally with some modification of stability class to compensate for the effects of roughness and source elevation. There is now available a considerable body of measurements from Brookhaven, NY, St. Louis, MO, and from Jülich and Karlsruhe in Germany pertaining to elevated sources and rough terrain. The diagrams shown in Fig. 10.5 are based on a summary of the German data by Geiss et al. (1981).

The diagrams are quite simple to use. One first determines the Pasquill stability class. This is best done on the basis of observations of σ_α if these are available. Otherwise one must estimate the class from the wind speed and the intensity of insolation, as deduced from the cloudiness using the Table 10.2. In no case should σ_z exceed the height of the mixed layer if the latter is known. One can then calculate the concentration at the desired location using either (10.35) or (10.36). Note that $\overline{u}$ in this equation refers to the wind at the equivalent height of the plume, not the wind at 5 meters that is used to determine the stability class. If $\overline{u}$ is not observed, it can be estimated from a power law or the Monin–Obukhov profile equation.

For distances smaller than 100 m, it should be assumed that σ_y and σ_z expand linearly. When using (10.22) with elevated sources it is preferable to estimate σ_α from σ_y at $x = 100$ m rather than from Table 10.1.

Perhaps the single most important question to be answered is the location and amount of the maximum concentration at the surface. It can easily be shown that if the source is at ground level, the maximum concentration is located at the source. If the source is elevated, then the surface location of the maximum expected value of the concentration must lie along the centerline somewhere downwind of the source. At the location of the source, σ_z is zero, and therefore the value of $H(z)$ is zero everywhere except at the source itself. Also, if one goes an infinite distance away in the downwind x-direction, σ_y becomes infinite and the value of the concentration becomes zero. Thus there must be at least one point between these limits where the concentration has a maximum value.

A solution is easily reached if we assume that σ_y is some constant a times σ_z, an assumption that is pretty well supported by observations. If we make this substitution in (10.37), and set $z = 0$, we then find the condition for maximum by differentiating χ with respect to x and setting the result equal to 0. The location of the point of maximum concentration is where σ_z is equal to seven-tenths of the effective height of the source, and the resulting value of the maximum concentration is

$$\chi_{\max} = \frac{2Q}{\pi h_e^2 e \overline{u}} \frac{\sigma_z}{\sigma_y} . \tag{10.40}$$

Users of charts such as Fig. 10.5 should not confuse the ease of getting estimates with the degree of reliability of the estimates. Probably the largest uncertainty stems from errors of estimating the stability class from surface observations. There are also large differences in the observed dispersion rates at different sites where data sets have been collected. For example, the German observation sites show a great deal of horizontal meandering of the plume under stable conditions. This phenomenon appears to be comprised of nonturbulent waves of unknown origin. Whatever is the source and nature of these motions, they cause a lateral spreading or fanning of the plume which, in the German data, causes σ_y to increase with increasing stability. These effects do not appear in the original Pasquill–Gifford diagrams.

One must also keep in mind that all data sets are limited in number because of the great cost of the experiments. Legislative requirements concern themselves with the maximum concentrations at the ground. These are usually associated with the most unstable stability parameters. However, unstable situations lead to enormous variations in the locations and magnitudes of one-hour maximum averages, so that the conclusions to be drawn from a limited number of experiments are subject to large quantitative uncertainties.

10.9 Estimations Based on Taylor's Equation

It is useful to define a function $f_v(T)$ such that

$$f_v(T) \equiv \frac{1}{\sigma_v^2} \int_0^\infty F_v(n) \frac{\sin^2(\pi nT)}{(\pi nT)^2} dn \tag{10.41}$$

or

$$\sigma_y = \sigma_v T \, f_v(T) = \sigma_\alpha X \, f_v(T) \tag{10.42}$$

after making the approximation that σ_y, at distance X from the source, is the same as that of the cloud of particles that left the source T seconds earlier. It has also been assumed that the Lagrangian variance of v is the same as the Eulerian variance, an assumption that is warranted when the turbulence is homogeneous and stationary. The meaning of $f_v(T)$ is then easily seen from Fig. 10.6 to be ratio of AB, the actual plume width at X, and AC, the width of projection of the tangents to the plume edge at the initial instant of release from the source. In the next sections we shall make use of observational determinations of this function.

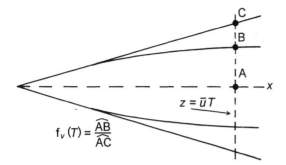

Fig. 10.6 Geometrical interpretation of the function $f_v(T)$

The problem of estimating σ_y or σ_z can be solved by finding the form of the function $f_v(T)$ defined in (10.42), and an analogously defined function for σ_z. We apply similarity principles, so that we expect that if T is measured in units of the Lagrangian time scale T_L, then $f_v(T/T_L)$ should be a universal function. Thus

$$\sigma_y = \sigma_v T f_v\left(\frac{T}{T_L}\right) = \sigma_\alpha X f_v\left(\frac{T}{T_L}\right) . \tag{10.43}$$

The parameters σ_v and σ_w can be determined using (8.22) or (8.23) and (8.18) or (8.20), respectively. (Note that (8.18) applies below $0.017h$ and (8.20) above.)

The form of the function is dependent on the form of the Lagrangian autocorrelation function. Observations are very difficult and therefore not very numerous. Most of the available observations have come from observing the motions of *tetroons*. These are balloons of tetrahedral shape, made of non-stretchable mylar plastic material, and filled with a buoyant gas mixture to a slight superpressure in such a way as to be neutrally buoyant. Ordinary balloons are not suitable because they tend to develop buoyancy when heated by sunlight.

The limited number of observations have been combined with hypotheses concerning the form of the function. We know, for example, that the value of the autocorrelation is 1 at zero time lag, and tends toward zero for large time separations. Further, its integral over all time is the integral time scale. The function

$$R(\xi) = \exp\left(\frac{-\xi}{T_L}\right) \tag{10.44}$$

satisfies these requirements as does also the following function used by Phillips and Panofsky (1982):

$$R(\xi) = \left(\frac{1+\xi}{T_L}\right)^{-2} . \tag{10.45}$$

Both functions are similar in form and are pictured in Fig. 10.7.

The Eulerian autocorrelation is much easier to measure because it depends on observations made at a fixed location. It too has similar characteristics of behavior,

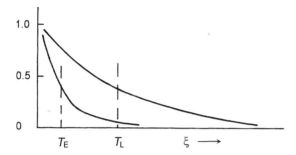

Fig. 10.7 Idealized Lagrangian and Eulerian autocorrelation functions

except that the time scale T_E is normally smaller than T_L. It is generally presumed that except for the time scales, both functions are the same. Thus, if we knew the ratio of T_E to T_L we could use observations of the Eulerian autocorrelation to deduce the form of the function $f(T/T_L)$. This ratio is generally given the symbol β.

The Eulerian time scale can be estimated by the use of Taylor's (frozen eddy) hypothesis by looking at the dominant eddy size l. The Eulerian time scale may be characterized as the reciprocal of the repetition rate of eddies of this size, which is proportional to $l/\overline{u}$. The Lagrangian time scale in a corresponding way is the repetition rate of the rotation of a particle's motion around the eddy, which is l/σ_v. The ratio is

$$\beta \equiv \frac{T_L}{T_E} = C\frac{\overline{u}}{\sigma_v} = \frac{C}{i} = \frac{C}{\sigma_\alpha} \tag{10.46}$$

in which i is called the intensity of turbulence (ratio of σ_v to $\overline{u}$). Theoretical considerations suggest a value for the coefficient

$$C = \frac{\sqrt{\pi}}{4} = 0.44 \tag{10.47}$$

or 25 if σ_α is measured in degrees. Observed values scatter over a large range, and suggest an average value of 0.7 (see Hanna (1981)).

Estimates of T_L at various heights and degrees of stability have been proposed by Draxler (1976) and Hanna (1981). Draxler's estimates have been converted from the time intervals he has given that are required for the autocorrelation to fall to a value of 0.5. These estimates of T_L are listed in the Table 10.3. Based on an analysis of tetroon flights, all in the daytime, Hanna suggested

$$T_L = \frac{0.17h}{\sigma_{u,v,w}} \tag{10.48}$$

for the following ranges of z:

Table 10.3 Draxler's values of T_L (seconds)

	Surface sources		Elevated sources	
	Stable	Unstable	Stable	Unstable
Lateral	60	60	200	200
Vertical	10	20	20	100

for u, v $0 \le z \le h$
for w $100\,\text{m} \le z \le h$

where h is the depth of the mixed layer. Below 100 m the vertical components of the eddy motions are constrained by the ground, whereas the horizontal components are not. In this region Reid (1976) has suggested

$$T_{LW} = 0.6z\beta/u_* , \quad z < 100\,\text{m} . \tag{10.49}$$

The form of the universal function $f_v(T/T_L)$ has been proposed by Draxler and by Phillips and Panofsky. Of the two, the latter is preferable since it is based on the form of the autocorrelation (10.45) which has the correct limiting behavior at short and long distances from the source;

$$f_v\left(\frac{T}{T_L}\right) = 1.41 \left[\frac{T_L}{T} - \left(\frac{T_L}{T}\right)^2 \ln\left(1 + \frac{T}{T_L}\right)\right]^{1/2} . \tag{10.50}$$

These results are seen to be quite tentative and leave a great many uncertainties to be resolved by further experiments. However, they probably constitute the best foothold that we now have for predicting the concentration downwind of a continuous source under given meteorological input conditions.

10.10 Monte Carlo Models

It is possible to simulate Taylor's equation without making restrictive assumptions about homogeneity and stationarity. This is done by studying the statistics of a large number of particles, each of which has moved stochastically from the source for a given time. The motion of each particle is independent of the motion of other particles, although it must at all times behave as though it was drawn at random from a population with known statistical properties.

The motion of individual particles is carried out in small time steps over the duration of the passage from the source to its final position after a time T. Initially the motion consists of the mean air velocity at the source plus a deviation that is a random member of a Gaussian velocity distribution with standard deviations in the different directions appropriate to its height and stability. At each new time step, the velocity is redetermined by application of the equation

$$v'(t + \xi) = R(\xi)\, v'(t) + v* \,, \tag{10.51}$$

in which $R(\xi)$ is the Lagrangian autocorrelation for a time lag equal to the time step ξ, and v^* is a randomly selected velocity from a Gaussian population with a variance equal to $1 - R^2(\xi)$ times the variance of v'. This scheme can be carried out for each of the velocity components independently. The effect of this equation is that each particle gradually loses its memory of the past but maintains its status as a random member of the population of the environment in which it finds itself.

For $R(\xi)$ the function (10.44) has generally been used with ξ equal to the time step. The function (10.45) would probably be more realistic, but seems not to have been used. In either case, it is necessary to estimate the Lagrangian time scale. For this purpose the equations in the preceding section can be used.

In the case of actual smoke plumes, successive particles leaving the source do not move independently of each other. Their initial velocity is determined by a succession of eddies of different sizes that are carried past the source by the mean wind. The effect of this procession can be simulated by determining the initial velocity components by an equation similar to (10.51) using the Eulerian autocorrelation in place of the Lagrangian autocorrelation. This is easily done by calculating β from (10.46). If one only wishes to determine long-term average concentrations, this refinement is unnecessary. It is useful, however, to be able to simulate actual realizations over time periods of various durations, typically an hour, to get some estimate of the uncertainty of simple Gaussian plume models.

On the diskette included with this publication is a Monte Carlo simulation based on the principles described in this chapter. A description of this model is given in Appendix E. Although the model incorporates the best information available at the present time, it has not been tested for engineering purposes. However, running it under a variety of input conditions gives a striking verification of the soundness of the principles discussed in this chapter. It is suggested that the student make a number of repeated runs under A-type stability conditions to observe the enormous variability of one-hour averages, and the corresponding difficulties of setting legislative standards for emissions under these conditions.

10.11 Instantaneous Point Sources

The expansion of a cloud, or cluster of particles leaving a source at the same time, is quite different in nature from the average concentration resulting from the continuous emission from a point source. In the latter case, each particle or small cluster leaving the source can be considered to be moving independently of the others, and, at least initially, the entire turbulent spectrum contributes to the expansion of the plume.

The individual members of a cluster or cloud of particles do not move independently of each other, but are subject to the two-point spatial correlations as functions of their separation. As we look at the effect of various components of the turbulent spectrum, only the eddies that are close to the size of the cloud are effective in expanding it. Eddies that are much smaller than the cluster mix it from within but have little effect on its size. Eddies that are much larger than the cluster simply move the cluster as a whole but do not expand it.

Panofsky (1984, p. 253) has pointed out that the effective eddies range in size from about 1/10 as large as the cloud to about 10 times the cloud size. Thus as one looks at the logarithmic spectrum, the effective eddy range constitutes a window of constant width centered at a point that moves through the spectrum starting at the high frequency end (when the cloud is very small) and moving left toward larger and larger sizes as the cloud expands. The rate of expansion at any stage is proportional to the average height of the logarithmic spectrum near the center of the window.

The size of the cluster is easily represented by the root mean square separation of the individual particles. The square of this is denoted by $\overline{\Delta Y^2}$. Initially the growth of the cluster is proportional to the average distance between particles; if the cluster size is actually zero initially, it cannot grow, since all the particles have the same velocity. Thus initially

$$\overline{\Delta Y^2} = \overline{\Delta Y_0^2} + \overline{(v_2 - v_1)^2}\, T^2 \; . \tag{10.52}$$

As the size grows larger, the rate of growth becomes independent of the separation of the particles and depends only on the spectral energy in the range of the window. It has been estimated that only about one ninth of the turbulent energy is effective at any one time in spreading the cluster. When the size of the cluster embraces eddies in the inertia subrange, the spectral density increases as the cluster grows and the window moves from right to left through the spectrum. In the inertia subrange, the rate of increase in the mean separation of the particles depends only on the rate of dissipation and the time. Dimensional analysis then demands

$$\overline{\Delta Y^2} = a\varepsilon T^3 \; . \tag{10.53}$$

As the window moves out of the inertia subrange into the energy-containing range, the logarithmic spectrum stops growing, and the rate of growth of the cluster becomes constant. Accordingly, is a linear function of T and the rate of growth in this central stage is proportional to the size itself. Finally, as the size of the cluster increases still further, the spectrum decreases and the rate of increase slows still further. Each particle moves independently of all the others, and the rate of growth becomes proportional to the square root of the time.

10.12 Problems

1. Suppose the concentration χ is independent of height between two plates, distance D apart, and zero elsewhere. Determine the function H at every height.

2. Suppose a chimney puts out 10 kg of sulfur dioxide per second. (This is not unusual for a power plant that has no scrubbers.) The plume is confined by an inversion 500 m above the ground and by valley walls that are 2 km apart horizontally. The wind is $5\,\mathrm{m\,s^{-1}}$ and is parallel to the valley walls. What is the concentration of sulfur dioxide 20 km downwind of the source? Express the answer in units of parts per million parts of air by mass (kilograms of sulfur dioxide per million kilograms of air).

3. Derive the equation (10.40) for the maximum concentration at the surface.

4. Show that (10.49) has the correct limiting behavior at small and large values of T/T_{L}. (See (10.21) and (10.24).)

5. Show that when the size of an expanding cluster or cloud corresponds to that of eddies in the inertia subrange, the cloud expands in proportion to $\varepsilon^{1/2}$ and time to the 3/2 power.

6. Explain why the average concentration of a plume resulting from a continuous point source is different from that within a cloud released at an instant of time, after it has traveled for the same length of time after its release (or the same distance downwind of its source).

Appendix A.
Derivation of the Turbulent Energy Equations

In Chap. 4, the energy equations of the atmosphere were discussed in a general way so as to make the important concepts clear without being burdened with details. In this appendix, we go through the derivation of the equations for various kinds and subdivisions of energy that enable one to describe the processes taking place in a turbulent portion of the atmosphere.

For the purposes of this appendix, we shall use p_{ki} and τ_{ki} exclusively for the total molecular stress and the viscous stress, and we shall use the notation T_{ki} to denote the Reynolds stresses.

A.1 Equations for the Instantaneous Energy

We begin with the statement of conservation of energy, often referred to as the first law of thermodynamics. This law may be stated as follows: the rate of change of the total energy of a system must equal the rate at which work is done on its boundary plus the rate at which heat enters the system through its boundary. This statement is a broadened version of the restricted principle of thermodynamics, since it deals with the rate at which these processes take place.

The system we deal with initially will consist of an infinitesimal parcel of fluid, and the rates of change referred to above apply always to the same parcel. Thus it is a closed system. Later when we consider the mean state of a turbulent system, we must redefine the system, which will inevitably be an open one, and we shall see how this statement of energy conservation must be modified.

The first task is to derive an expression for the rate at which work is done on the boundary of a parcel. For this purpose we look at the small parallelepiped pictured in Fig. 2.2. We first find an expression for the work done by the surrounding fluid on the lower surface of the parcel. As we have seen before, the total force exerted on the underside (i.e. outside) is the vector $-p_{3i}\mathrm{d}x_1\mathrm{d}x_2$, and the rate at which work is done by this force is the dot product of the force and the velocity vector u_i, or

$$-p_{3i}u_i\,\mathrm{d}x_1\,\mathrm{d}x_2\ . \tag{A.1}$$

We now move to the upper surface of the box, taking into consideration the changes in both the stress and the velocity as a result of the separation of the two surfaces by dz:

$$\left(p_{3i} + \frac{\partial p_{3i}}{\partial x_3} dx_3\right)\left(u_i + \frac{\partial u_i}{\partial x_3} dx_3\right) dx_1 dx_2 \,. \tag{A.2}$$

We add (A.1) and (A.2) together, expand the products, and neglect terms in dx_3^2, and then divide by the mass of the fluid inside box. The result is

$$\frac{1}{\rho}\left(u_i \frac{\partial p_{3i}}{\partial x_3} + p_{3i}\frac{\partial u_i}{\partial x_3}\right) = \frac{1}{\rho}\frac{\partial\left(p_{3i}u_i\right)}{\partial x_3} \,. \tag{A.3}$$

It is easily seen that with the addition of the other two faces, this expression becomes

$$\text{Work done per unit time, per unit mass} = \frac{1}{\rho}\frac{\partial\left(p_{ki}u_i\right)}{\partial x_k} \,. \tag{A.4}$$

As we have done before, we find it helpful to separate the stress into its two parts, the pressure p and the viscous stress τ_{ki}.

$$\frac{1}{\rho}\frac{\partial\left(p_{ki}u_i\right)}{\partial x_k} = -\frac{1}{\rho}\frac{\partial p u_i}{\partial x_i} + \frac{1}{\rho}\frac{\partial\left(\tau_{ki}u_i\right)}{\partial x_k} \,. \tag{A.5}$$

The first of these terms comprises the rate at which work is done by the pressure. To put this into better perspective, we expand the derivative of the product and employ the equation of continuity in the form (α being the volume occupied by a unit mass):

$$\frac{\partial u_i}{\partial x_i} = \frac{1}{\alpha}\frac{d\alpha}{dt} \tag{A.6}$$

so as to obtain

$$-\frac{1}{\rho}\frac{\partial p u_i}{\partial x_i} = -\frac{u_i}{\rho}\frac{\partial p}{\partial x_i} - p\frac{d\alpha}{dt} \,. \tag{A.7}$$

It is seen from this, that the total rate at which work is done by the pressure on the boundary of a unit mass consists of two parts: (1) the work done by the pressure gradient force (which appears in the Navier–Stokes equations) and (2) the thermodynamic work of expansion that appears in the first law of thermodynamics. As long as one confines attention to systems that are at rest and in equilibrium, the first type of work and the role it plays are never considered.

In a similar way we turn to the viscous terms in (A.5). Expansion of these leads to

$$\frac{1}{\rho}\frac{\partial\left(\tau_{ki}u_i\right)}{\partial x_k} = \frac{u_i}{\rho}\frac{\partial \tau_{ki}}{\partial x_k} + \frac{\tau_{ki}}{\rho}\frac{\partial u_i}{\partial x_k} \,. \tag{A.8}$$

By comparison with (A.7), we interpret the first set of terms as the rate at which work is done by the stress-gradient forces in the Navier–Stokes equations, the so-called frictional forces that are effective in changing the momentum of fluid elements.

The second set of terms in (A.8), which will henceforth be denoted by ε_T, may be expanded making use of the symmetry of the stress and the definition of the pure deformation rate d_{ik} in (2.7)

$$\varepsilon_T \equiv \frac{\tau_{ki}}{\rho}\frac{\partial u_i}{\partial x_k} = \frac{\tau_{ki}}{\rho}\frac{1}{2}\left(\frac{\partial u_i}{\partial x_k} + \frac{\partial u_k}{\partial x_i}\right) = \frac{\tau_{ki}}{\rho}\frac{1}{2}\left(d_{ik} + \frac{2}{3}\frac{\partial u_j}{\partial x_j}\delta_{ik}\right) \tag{A.9}$$

whence from the relation between viscous stress and d_{ik}, we find

$$\varepsilon_T = \frac{v}{2}(d_{ik})^2 . \tag{A.10}$$

Thus ε_T is a positive definite form, and it is always positive if the kinematic viscosity is positive. We shall see later that this is guaranteed by the second law of thermodynamics. As will be seen below, this set of terms represents the rate of dissipation of mechanical energy into thermodynamic internal energy.

We now turn to the equation that expresses the conservation of energy. We shall assume that the total energy per unit mass consists of kinetic energy $u_i^2/2$, gravitational potential energy Φ, and the thermodynamic internal energy $c_v T$. The statement of energy conservation may be written

$$\frac{d}{dt}\left(\frac{u_i^2}{2} + \Phi + c_v T\right) = -\frac{u_i}{\rho}\frac{\partial p}{\partial x_i} - p\frac{d\alpha}{dt} + \frac{u_i}{\rho}\frac{\partial \tau_{ki}}{\partial x_i} + \varepsilon_T - \frac{1}{\rho}\frac{\partial h_k}{\partial x_k}$$

$$= -\frac{1}{\rho}\frac{\partial p_{ki}}{\partial x_k} - \frac{1}{\rho}\frac{\partial h_k}{\partial x_k} . \tag{A.11}$$

The last term in this equation contains the rate of convergence of the heat flux h_k. In this case the flux includes all forms of heat transfer applicable to a closed system, viz. radiation and conduction. When this term is integrated over the mass of a finite system, making use of Gauss's divergence theorem, the result is the net rate at which heat enters the mass across its bounding surface. Equation (A.11) is seen to be a generalized statement of the first law of thermodynamics. The classical form of the first law is a special case of this equation which prevails when the system is at rest, the pressure is spatially uniform, and the deformation rate is infinitesimal.

It is instructive to deduce equations for each of the individual forms of energy. The simplest of these is the potential energy equation

$$\frac{d\Phi}{dt} = gw . \tag{A.12}$$

The kinetic energy equation is obtained by multiplying the Navier–Stokes equations by u_i and summing over i, noting

$$u_i\frac{du_i}{dt} \equiv \frac{d}{dt}\frac{u_i^2}{2} , \tag{A.13}$$

$$\frac{d}{dt}\frac{u_i^2}{2} = -gw - \frac{u_i}{\rho}\frac{\partial p}{\partial x_i} + \frac{u_i}{\rho}\frac{\partial \tau_{ki}}{\partial x_k} = -gw + p\frac{d\alpha}{dt} - \varepsilon_T + \frac{1}{\rho}\frac{\partial (p_{ki}u_i)}{\partial x_k} . \quad (A.14)$$

Finally we obtain an equation for the thermodynamic internal energy by subtracting both (A.12) and (A.14) from (A.11).

$$\frac{dc_v T}{dt} = -p\frac{d\alpha}{dt} + \varepsilon_T - \frac{1}{\rho}\frac{\partial h_k}{\partial x_k} . \quad (A.15)$$

The last equation is the familiar limited statement of the first law except for the dissipation, which is usually left out of elementary thermodynamic considerations by assuming that all changes proceed very slowly.

It is useful to look at the set of three equations for the separate forms of energy (A.12), (A.13), and (A.14), together with (A.11) for the total energy. For this purpose, we rewrite all equations in a form that refers to unit volume, using the equation of continuity, in the manner discussed in the sequence of equations (1.12) to (1.15) on Sects. 1.7–1.8.

$$\frac{\partial}{\partial t}\left(c_v\rho T + \rho\Phi + \frac{\rho u_i^2}{2}\right)$$

$$= -\frac{\partial}{\partial x_k}\left(c_v\rho u_k T + \rho u_k\Phi + \frac{\rho u_k u_i^2}{2}\right) + \frac{\partial (p_{ki}u_i)}{\partial x_k} - \frac{\partial h_k}{\partial x_k} , \quad (A.16a)$$

$$\frac{\partial \rho\Phi}{\partial t} = -\frac{\partial \rho u_k\Phi}{\partial x_k} + g\rho w , \quad (A.16b)$$

$$\frac{\partial}{\partial t}\frac{\rho u_i^2}{2} = -\frac{\partial}{\partial x_k}\left(\frac{\rho u_k u_i^2}{2}\right) - g\rho w + \rho p\frac{d\alpha}{dt} - \rho\varepsilon_T + \frac{\partial (p_{ki}u_i)}{\partial x_k} , \quad (A.16c)$$

$$\frac{\partial (c_v\rho T)}{\partial t} = -\frac{\partial (c_v\rho u_k T)}{\partial x_k} - \rho p\frac{d\alpha}{dt} + \rho\varepsilon_T - \frac{\partial h_k}{\partial x_k} . \quad (A.16d)$$

The term on the left side of each equation is the change per unit volume and per unit time of the various forms of energy at a fixed point. The first term on the right side represents the convergence of the flux of energy carried in by the fluid motion, as discussed in the first chapter (Sect. 1.7). The remaining terms on the right side represent the sources of each form of energy. For the total energy, the sources can only come from outside the boundaries of the volume. These consist of the work done by the stresses, which are seen to affect the kinetic energy, and the heat, which affects the thermodynamic internal energy. The potential energy source consists only of vertical motion through the term $g\rho w$, and this same term appears in the kinetic energy with opposite sign. We can therefore regard this term as the rate of transformation of one form of energy into the other, and it may be in either direction depending on its sign. Similarly the two terms in

$$-\rho p\frac{d\alpha}{dt} + \rho\varepsilon_T \quad (A.17)$$

represent sources of thermodynamic internal energy and a simultaneous loss of kinetic energy per unit volume and per unit time. We see these terms as representing the rate of transformation of kinetic energy into heat. We note that none of the transformation terms appear in the equation for the total energy; their role is to change the amounts of the individual forms without changing the total in any way.

A.2 The Equation of Mean Internal Energy

We begin with equation (A.14) and apply the identity

$$p\frac{\mathrm{d}\alpha}{\mathrm{d}T} \equiv \frac{\mathrm{d}(p\alpha)}{\mathrm{d}t} - \alpha\frac{\mathrm{d}p}{\mathrm{d}t} = \frac{\mathrm{d}(RT)}{\mathrm{d}t} - \alpha\frac{\mathrm{d}p}{\mathrm{d}t} \tag{A.18}$$

to get the alternate form of the generalized first law,

$$\rho\frac{\mathrm{d}c_p T}{\mathrm{d}t} - \frac{\mathrm{d}p}{\mathrm{d}t} = \rho\varepsilon_T - \frac{\partial h_k}{\partial x_k} . \tag{A.19}$$

Applying the equation of continuity in the now familiar way so as to move to an open-system configuration, and expanding $\mathrm{d}p/\mathrm{d}t$ we obtain

$$\frac{\partial\left(c_p\rho T\right)}{\partial t} + \frac{\partial\left(c_p\rho u_k T\right)}{\partial x_k} - \frac{\partial p}{\partial t} - u_k\frac{\partial p}{\partial x_k} = \rho\varepsilon_T - \frac{\partial h_k}{\partial x_k} . \tag{A.20}$$

We now average this equation using the Reynolds Postulates;

$$\frac{\partial c_p\overline{\rho}\overline{T}}{\partial t} + \frac{\partial\left(c_p\overline{\rho}\ \overline{u_k T}\right)}{\partial x_k} - \frac{\partial\overline{p}}{\partial t} - \overline{u_k}\frac{\partial\overline{p}}{\partial x_k} = \rho\varepsilon_T - \frac{\partial\left(\overline{h_k} + c_p\overline{\rho}\overline{u_k' T'}\right)}{\partial x_k} . \tag{A.21}$$

In this rewritten result, the eddy flux of heat that comes out of the second term on the left, is now placed on the right side beside the mean heat flux $\overline{h_k}$. We have also neglected a small term involving the pressure fluctuation since we have seen that it is small compared to the temperature fluctuations.

To put this in more familiar terms, multiply the mean equation of continuity (1.10) by $c_p\overline{T}$ and subtract it from (A.21) term-by-term. We also note

$$\frac{\mathrm{D}\overline{p}}{\mathrm{D}t} = \overline{\rho}\ \overline{\alpha}\frac{\mathrm{D}\overline{p}}{\mathrm{D}t} = \frac{\mathrm{D}R\overline{T}}{\mathrm{D}t} - \overline{p}\frac{\mathrm{D}\overline{\alpha}}{\mathrm{D}t} . \tag{A.22}$$

The final result is then the equation

$$\frac{\mathrm{D}c_v\overline{T}}{\mathrm{D}t} + \overline{p}\frac{\mathrm{D}\overline{\alpha}}{\mathrm{D}t} = (\varepsilon_M + \varepsilon) - \frac{1}{\overline{\rho}}\frac{\partial\left(\overline{h_k} + c_p\overline{\rho}\overline{u_k' T'}\right)}{\partial x_k} . \tag{A.23}$$

A.3 The Mean Total Kinetic Energy Equation

We begin with a slightly altered form of (A.16c);

$$\frac{\partial}{\partial t}\frac{\rho u_i^2}{2} + \frac{\partial}{\partial x_k}\frac{\rho u_k u_i^2}{2} = -g\rho w + p\frac{\partial u_i}{\partial x_i} - \tau_{ki}\frac{\partial u_i}{\partial x_k} + \frac{\partial (p_{ki}u_i)}{\partial x_k} \tag{A.24}$$

which is ready to be averaged using the Reynolds postulates. The mean kinetic energy is then seen to consist of two parts:

$$\overline{E}_T \equiv \frac{\overline{u_i^2}}{2} = \frac{\overline{u}_i^2}{2} + \frac{\overline{u_i'^2}}{2} = E_M + \text{TKE} . \tag{A.25}$$

The averaged equation then is

$$\frac{\partial \overline{\rho}(E_M + \text{TKE})}{\partial t} + \frac{\partial [\overline{\rho u}_k (E_M + \text{TKE})]}{\partial x_k}$$

$$= -g\rho \overline{w} - g\overline{\rho' w'} + \overline{p}\frac{\partial \overline{u}_i}{\partial x_i} - \overline{\rho}\varepsilon_M - \overline{\rho}\varepsilon + \frac{\partial}{\partial x_k}\left(\overline{p}_{ki}\overline{u}_i + \overline{p'_{ki}u'_i} - \overline{\rho\frac{u'_k u_i'^2}{2}}\right) . \tag{A.26}$$

Again in this equation a small term involving pressure fluctuations has been omitted. We revert to a change following the mean motion by multiplying the mean equation of continuity by $(E_M + \text{TKE})$ term-by-term and subtracting. We also make some changes as before to obtain the result

$$\frac{D(E_M + \text{TKE})}{Dt}$$

$$= -g\overline{w} - \frac{g}{\overline{\rho}}\overline{\rho' w'} + \overline{p}\frac{D\overline{\alpha}}{Dt} - \varepsilon_M - \varepsilon + \frac{1}{\overline{\rho}}\frac{\partial}{\partial x_k}\left(\overline{p}_{ki}\overline{u}_i + \overline{p'_{ki}u'_i} - \overline{\rho\frac{u'_k u_i'^2}{2}}\right) . \tag{A.27}$$

A.4 The Equation for the Energy of Mean Motion

We begin with a slightly changed form of the averaged Navier–Stokes equation (2.13);

$$\frac{D\overline{u}_i}{Dt} = -g\delta_{3i} - \frac{1}{\overline{\rho}}\frac{\partial \overline{p}}{\partial x_i} + \frac{1}{\overline{\rho}}\frac{\partial}{\partial x_k}\left(\overline{\tau}_{ki} - \overline{\rho u'_k u'_i}\right) . \tag{A.28}$$

We get an equation for E_M by multiplying through by $\overline{u}_i$ and summing over i. The result is

$$\frac{DE_M}{Dt} = -g\overline{w} - \frac{\overline{u}_i}{\overline{\rho}}\frac{\partial \overline{p}}{\partial x_i} + \frac{\overline{u}_i}{\overline{\rho}}\frac{\partial}{\partial x_k}\left(\overline{\tau}_{ki} - \overline{\rho u'_k u'_i}\right) . \tag{A.29}$$

We note the following:

$$\bar{u}_i \frac{\partial \bar{p}}{\partial x_i} \equiv \frac{\partial}{\partial x_i}(\bar{p}\,\bar{u}_i) - \bar{p}\frac{\partial \bar{u}_i}{\partial x_i} \tag{A.30}$$

and other identities like it. Also, we have, by definition

$$\varepsilon_M = \bar{\tau}_{ik}\frac{\partial \bar{u}_i}{\partial x_k}\ .$$

Use of these relations gives

$$\frac{DE_M}{Dt} = -g\bar{w} + \frac{\bar{p}}{\bar{\rho}}\frac{\partial \bar{u}_i}{\partial x_i} - \varepsilon_M + \overline{u_k' u_i'}\frac{\partial \bar{u}_i}{\partial x_k}$$
$$- \frac{1}{\bar{\rho}}\frac{\partial \bar{p}\,\bar{u}_i}{\partial x_i} + \frac{1}{\bar{\rho}}\frac{\partial}{\partial x_k}\left[\left(\bar{\tau}_{ki} - \overline{\rho u_k' u_i'}\right)\bar{u}_i\right]\ . \tag{A.31}$$

Again we use the equation of continuity

$$\frac{\partial \bar{u}_i}{\partial x_i} = -\frac{1}{\bar{\rho}}\frac{D\bar{\rho}}{Dt} = +\frac{1}{\bar{\alpha}}\frac{D\bar{\alpha}}{Dt} \tag{A.32}$$

and combine the remaining pressure term with the viscosity terms to obtain the final result

$$\frac{DE_M}{Dt} = -g\bar{w} + \bar{p}\frac{D\bar{\alpha}}{Dt} - \varepsilon_M - M + X_M\ , \tag{A.33}$$

in which

$$M \equiv -\overline{u_k' u_i'}\,\frac{\partial \bar{u}_i}{\partial x_k} \tag{A.34}$$

and

$$X_M \equiv \frac{1}{\bar{\rho}}\frac{\partial}{\partial x_k}\left[\left(\bar{p}_{ki} - \overline{\rho u_k' u_i'}\right)\bar{u}_i\right]\ . \tag{A.35}$$

A.5 The Turbulent Kinetic Energy Equation

The equation for TKE follows immediately by subtracting the equation for the energy of the mean motion E_M, which is found in (A.33), from the equation for the mean total kinetic energy (E_M + TKE), which appears in (A.27). The result is

$$\frac{D(TKE)}{Dt} = -\frac{g}{\bar{\rho}}\overline{\rho' w'} - \varepsilon + M + X_E \tag{A.36}$$

in which X_E is defined as

$$X_E \equiv \frac{1}{\bar{\rho}}\frac{\partial}{\partial x_k}\left(\overline{p'_{ki} u'_i} + \overline{\rho u_k' u_i'}\,\bar{u}_i - \bar{\rho}\frac{\overline{u_k' u_i'^2}}{2}\right)\ . \tag{A.37}$$

All of these equations are thoroughly discussed in Chap. 4.

Appendix B.
Dimensional Analysis and Scaling Principles

Some of the most important breakthroughs of science have come from the shrewd application of scaling principles. The study of turbulence provides several remarkable examples of the power of this method.

Physical quantities have many attributes, and the association of these quantities with their attributes gives valuable clues that can narrow down the number of possibilities for relating causes and effects. A vector cannot equal a scalar any more than apples can equal oranges. We have seen how Stokes was able to infer a complex relation between viscous stress and the velocity distribution by observing this principle.

Dimensional analysis concerns itself with the attributes of mass M, length L, time T, and temperature D. The main object of the analysis of dimensions is to reduce the number of variables that are involved in the relationship between two physical quantities. In so doing, it is possible to greatly reduce the number of observations that are needed to determine the relationship. In some cases, equations can be reduced to nondimensional form, and the relationship to a single curve involving pure numbers. Theories that accomplish this are called similarity theories.

Dimensional analysis has other applications of interest to students. Habitual checking of dimensions gives early warning signals of errors of memory, or errors of copying. Such errors are usually easily corrected if they are recognized. Lapses of memory in stressful situations such as examinations need not be fatal; methodical application of dimensional principles can enable forgotten relations to be reconstructed.

B.1 Checking Equations for Errors

Reconstructing an equation from memory entails obvious risks. Many errors can be detected and corrected if one habitually employs dimensional analysis. One might suppose for example that a student taking an examination must reconstruct the equation for the power P generated by a windmill and comes up with the result

$$P = 0.123 \ AV^2 \tag{B.1}$$

in which A stands for the area of the blades, and V stands for the wind speed. A quick check yields the following:

$$[P] = M\,L^2\,T^{-3} \tag{B.2}$$

where the brackets [ ] signify a dimensional statement. Likewise,

$$[A] = L^2 \tag{B.3}$$

$$[V] = L\,T^{-1} \tag{B.4}$$

$$[0.123] = M^0 L^0 T^0 D^0 = 1 \tag{B.5}$$

The last equation indicates a pure number lacking in dimensions.

The dimensions of each side of the presumed equation then produces the following test identity.

$$ML^2T^{-3} \equiv?\ \ L^2 \times L^2T^{-2}\,. \tag{B.6}$$

Identity here requires that the exponent of each individual dimension be the same on both sides. Obviously this requirement is not satisfied. The presumed equation cannot be correct! Note that even if the equation had balanced, the validity of the constant would not have been verified.

B.2 Inferring an Unknown Relationship

Frequently a guess is all that is needed to correct an error or to infer the form of an unknown relation, and the validity of the guess is easily determined by the dimensional test. Usually, however, a more methodical approach is required.

The first and most important step in this process is to identify the physical entities that are involved in the phenomenon or process under consideration. In the case of the windmill, the air density must be involved. Also obvious is the existence of a wind V. A little further thought reveals the likelihood that the width d or area A of the vanes should affect the result.

One proceeds by trying a relationship of the form

$$P = k\,\rho^a d^b V^c\,. \tag{B.7}$$

In this equation, k is a dimensionless constant the value of which cannot be determined by dimensional reasoning, and a, b, and c are numerical exponents to be determined. If it turns out that such a relationship is impossible, the presumed equation must be incorrect. Most likely the wrong choice of variables has been made. If it turns out that the hypothetical relationship is possible, there is still no guarantee that it is correct. However, it is always preferable to know that the presumed relationship could be right than to know that it cannot possibly be right.

We proceed methodically. Using (B.7) we write the corresponding dimensional equation

$$ML^2T^{-3} \equiv M^a L^{-3a} \times L^b \times L^c T^{-c} . \tag{B.8}$$

The identity sign indicates that the sum of the exponents of each dimension must separately be equal. It results in the following set of equations that must be satisfied.

$$a = 1 , \tag{B.9a}$$

$$-3a + b + c = 2 , \tag{B.9b}$$

$$-c = -3 . \tag{B.9c}$$

The solution is

$$a = 1 , \quad b = 2 , \quad c = 3 . \tag{B.10}$$

The existence of a unique solution implies that the relationship

$$P = k\rho d^2 V^3 \tag{B.11}$$

is the only one involving the chosen variables that can be correct. However, there is no guarantee that the original choice of variables was correct.

Final verification of the inferred relation as well as the determination of the constant k requires that experiments be performed. Dimensional analysis can be useful in showing the most efficient way to test the relation. If the relation is correct, one can define a nondimensional power function P^* by the equation

$$P^* \equiv \frac{P}{\rho d^2 V^3} \tag{B.12}$$

which, it is expected, will be found to be constant for any given design of windmill. As is frequently done in physics, one could give this number a name, such as the *Power Number*. Alternatively, one could describe P^* as a universal function of other dimensionless parameters of windmill design. The crucial experiments then reduce to making measurements of the power P under various combinations of d (different sizes of the same windmill design), air density, and wind speed to find if the value of P^* is really a constant and what its value is for that design.

B.3 Turkey Eggs, Anybody?

A number of years ago, my spouse returned from a visit to a farmer's market with what she expected would be a special surprise for the family – a bag of turkey eggs. She knew from experience that we like eggs boiled for 5 minutes. But turkey eggs weigh about twice as much as chicken eggs, and she realized that it would take longer than 5 minutes to boil them in the style to which we were accustomed.

There is more than one way to attack a problem like this. Some people would probably approach a solution by trial and error, hoping there are enough eggs in the bag to arrive at an acceptable answer, and probably sacrificing a few eggs to inexperience.

However, experience is the result of previously conducted experiments. Just because certain dimensions are changed should not require that one start from a state of ignorance. Two important things are known: (1) chicken eggs should be boiled for five minutes; and (2) the process of cooking an egg is governed by (4.18), which is repeated below:

$$\frac{\partial T}{\partial t} = \kappa \frac{\partial^2 T}{\partial x_k^2}$$

In most cases of physics, and particularly in atmospheric turbulence, it is not possible – nor is it necessary – to solve the equations that govern the important processes that we are dealing with. The unsolved equation can yield valuable information about how to design an efficient experiment (with the smallest number of variables) or, in the present case, about how to best use the experience that has already been accumulated.

The first step is to nondimensionalize the variables in a physically meaningful way. This procedure is called *scaling*. Dividing x_k by the distance from New York to Chicago would nondimensionalize it, but it would not be a useful scale. Considering the nature of the relation we are hoping to find, we should probably choose a length d that describes the size of the egg, say, its diameter. We then introduce a nondimensional set of distances $\xi_k \equiv x_k/d$. In a similar way we choose a time scale t_s that is identified with the phenomenon, say the time it takes to produce a properly cooked egg, and we let $\tau \equiv t/t_s$ stand for the scaled time. Finally, we must scale the temperature change by some temperature difference that we can identify with the properly cooked egg. A good choice would be the difference between the temperature T_e of the surrounding fluid and T_0, the temperature at the egg's center at the initial time. A useful scaled temperature variable would then be $\theta = (T - T_0)/(T_e - T_0)$.

The differential equation governing the process can now be rewritten using the scaled variables

$$\frac{\partial \theta}{\partial \tau} = \kappa^* \frac{\partial^2 \theta}{\partial \xi_k^2} \quad \text{in which} \quad \kappa^* \equiv \frac{\kappa t_s}{d^2}$$

is a dimensionless constant. Since all of the variables in the rewritten equation are dimensionless and scaled in terms of quantities identified with the process of egg cooking, the equation should be universally applicable to eggs of any size. We also observe that during the cooking process, the independent variables change from zero to a limit of unity, and that all terms of the equation, as well as the constant κ^* are of order unity. This fact is an important test of the validity of the scaling that was used. The value of the constant need not be determined. It is sufficient to know that its value is the same for a turkey egg as for a chicken egg. One then easily finds

$$(t_s)_{\text{turkey}} = \frac{(d^2)_{\text{turkey}}}{(d^2)_{\text{chicken}}} (t_s)_{\text{chicken}}$$

which for the given conditions yields a result of 7 minutes and 56 seconds. (This turned out to be just right!)

Inferences of relationships involving turbulence are often not clear because the variables involved are obscure, and testing is difficult or inconclusive. The principle has been suggested that dimensionless constants of valid relationships are usually of order unity or not too different from it. Constants that turn out to be very large or very small numbers tend to suggest that the selection of physical variables is incorrect. For example, the Charnock relation (3.7) for the roughness length over the sea might be questioned on this basis. The fact that deposition velocities (3.17), which include in their definition an implied dimensionless constant, vary over such a wide range of orders of magnitude is also strongly indicative of an incorrect formulation.

B.4 Problems

1. The first successful human-propelled airplane was designed and built by a meteorologist (viz. Dr Paul McCready), who understood the importance of dimensional analysis. Using such principles, determine the design constraints (total weight, wing area, etc.) required to minimize the ratio of the power needed for sustained flight to the total weight. Assume that the horizontal drag force is proportional to the lift (vertical force supplied by the wings). Would the requirements be met more easily by flying at high altitudes (low air density) or low altitudes? Would such an airplane be likely to set speed records?

2. The energy spectrum is defined as the turbulent kinetic energy (per unit mass) contained within a unit range of wavenumbers. (The wavenumber is the reciprocal of the wavelength.) According to the equilibrium theory of the turbulence spectrum (see Chap. 9), there is a large portion of the spectrum (called the inertial subrange) in which the spectrum depends only on the wavenumber and on the rate of dissipation of kinetic energy per unit mass. Show that, if this is so, then the spectrum is proportional to the minus five-thirds power of the wavenumber.

Appendix C.
Matching Theory and the PBL Resistance Laws

In Chap. 6 a model was presented for the distribution of wind for the entire layer in which the wind is disturbed from the geostrophic free-stream velocity. An important result of this model was a relation between the wind in the surface layer and the geostrophic free stream velocity. In this appendix, a more rigorous presentation of the theory of this layer is presented together with a derivation of these so-called resistance laws that is more general. As in Chap. 6, we assume that the mean flow is horizontally homogeneous and stationary.

Were it not for the stress at the boundary, the velocity throughout the planetary boundary layer (PBL) would everywhere be equal to the free-stream velocity. In the atmosphere, this velocity is the geostrophic velocity. The deviation from this velocity is called the *velocity defect*, and since it is caused by the stress at the surface it is expected that it should scale as $u^* = \sqrt{\tau_0/\rho}$, where τ_0 is the stress at the surface.

For homogeneous, stationary conditions, the equations of motion are

$$\overline{\rho} f \left(\overline{v} - v_\mathrm{g} \right) = -\frac{\mathrm{d}\tau_x}{\mathrm{d}z} \quad \text{and} \quad \overline{\rho} f \left(\overline{u} - u_\mathrm{g} \right) = \frac{\mathrm{d}\tau_y}{\mathrm{d}z} , \tag{C.1}$$

where

$$\tau_x \equiv -\overline{\rho w' u'} \quad \text{and} \quad \tau_y \equiv -\overline{\rho w' v'} , \tag{C.2}$$

and u_g and v_g are the geostrophic wind components. We now attempt to scale the velocity defects and the stress components in such a way that they are represented through the PBL by purely numerical variables of order unity. (We have previously noted that dimensionless constants vastly different from one are indicative of a physically incorrect formulation.) This is accomplished be scaling the velocity defects by u^* and the stresses by ρu^{*2}. Note that the stress components change with height, but u^*, having been defined by the stress at the surface is a constant with the dimensions of velocity. The result of the scaling is

$$\frac{\left(\overline{v} - v_\mathrm{g} \right)}{u^*} = -\frac{u_*}{f} \frac{\mathrm{d}}{\mathrm{d}z} \left(\frac{\tau_x}{\rho u_*^2} \right) ; \quad \frac{\left(\overline{u} - u_\mathrm{g} \right)}{u_*} = \frac{u_*}{f} \frac{\mathrm{d}}{\mathrm{d}z} \left(\frac{\tau_y}{\rho u_*^2} \right) . \tag{C.3}$$

The height remains to be scaled in order to completely nondimensionalize the variables. The equations (C.3) indicate that the stress decreases from its surface value to zero at heights of the order of u^*/f, which has a typical value of the order of 10^3 to 10^4 m. If we use this variable to scale a new nondimensional height Z,

$$Z = zu_*/f \; ; \tag{C.4}$$

then all the terms of the new dimensionless equations are of order unity;

$$\frac{(\bar{v} - v_g)}{u_*} = -\frac{d}{dZ}\left(\frac{\tau_x}{\bar{\rho}u_*^2}\right) \; ; \quad \frac{(\bar{u} - u_g)}{u_*} = \frac{d}{dZ}\left(\frac{\tau_y}{\bar{\rho}u_*^2}\right) \; . \tag{C.5}$$

One might conclude from this equation that the solution is a function of Z only. However, the solution must also depend on the boundary condition at the lower surface. Thus the solution must also depend on z_0, the surface roughness. In fact, close to the surface, the height scales that are involved in describing the velocity distribution are of the order z_0 rather than of order u_*/f. The ratio of the two length scales is called the *surface Rossby number*

$$\text{Ro} \equiv \frac{u_*}{f z_0} \; , \tag{C.6}$$

and this is of order 10^4–10^9, depending on the kind of surface and the latitude. We are faced with the fact that the solution depends on two different variables of vastly different orders of magnitude, namely Z of order unity, and Ro approaching an infinite order.

To clarify the concepts involved here, we consider the case in which the surface Rossby number approaches infinity. Throughout the bulk of the PBL the scaled velocity defects can be a function only of Z, for something of order unity cannot depend on something that approaches infinity in size. Such independence of the surface Rossby number is called *Rossby number similarity*. Close to the surface, the scaled velocity (not the velocity defect) must be a function of $\varsigma \equiv z/z_0$, for only in this way can the boundary condition be satisfied. Clearly this solution, which is also of order unity, must be independent of Z since the ratio of Z to z_0 is of the same order of magnitude as the surface Rossby number, which is approaching infinity.

We can summarize this situation in the following way. In the outer layer $(z \gg z_0)$,

$$\frac{(\bar{u} - u_g)}{u_*} = F(Z) \; ; \quad \frac{(\bar{v} - v_g)}{u_*} = G(Z) \tag{C.7}$$

while in the inner layer, close to the surface,

$$\frac{\bar{u}}{u_*} = f(\varsigma) \; ; \quad \frac{\bar{v}}{u_*} = 0 \; . \tag{C.8}$$

Since the ratio of length scales is approaching infinity, there must be a broad region in which z is large compared to z_0 and at the same time is small compared to u_*/f. Within this overlapping region, both solutions should be the same. Under what conditions can both (C.7) and (C.8) be identical? This requirement can be expressed in the following way:

$$f(\varsigma) - \frac{u_g}{u_*} \equiv F(Z) \; ; \quad -\frac{v_g}{u_*} \equiv G(Z) \tag{C.9}$$

where the identity symbols signify that the two sides must be identical over the entire overlapping region.

The required condition is found by differentiating both sides with respect to ς, noting $dZ/d\varsigma = fz_0/u_* = Z/\varsigma$. With primes denoting differentiation with respect to the argument, the result can be written

$$\varsigma f'(\varsigma) \equiv ZF'(Z) . \tag{C.10}$$

The left side is a function of ς only, while the right side is a function of Z only. The only way they can be equal for a wide range of heights is for each to be equal to a constant. We set the constant equal to $1/k$ where k is an undefined constant. The left side then gives

$$\frac{d\left(\overline{u}/u_*\right)}{d\ln\varsigma} = \frac{1}{k} \tag{C.11}$$

or

$$f(\varsigma) = \frac{\overline{u}}{u_*} = \frac{1}{k}\ln\frac{z}{z_0} \tag{C.12}$$

Thus we see that in the immediate vicinity of the boundary, the wind must be logarithmic, and we can identify k with the von Karman constant. Integration of the right sides gives the result

$$F(Z) = \frac{1}{k}\left(\ln\frac{zf}{u_*} + A\right) \; ; \quad G(Z) = \frac{B}{k} , \tag{C.13}$$

where A and B are undetermined constants of integration. These solutions may be substituted into (C.9) with the result

$$\frac{u_g}{u_*} = \frac{1}{k}\left[\ln\left(\frac{u_*}{fz_0}\right) - A\right] \; ; \quad \frac{v_g}{u_*} = -\frac{B}{k} . \tag{C.14}$$

These equations, originally derived by a different method by Kazanski and Monin, are usually called the resistance laws.

Under diabatic conditions, (C.7) must be broadened to include a stability parameter. For this,

$$\sigma \equiv \frac{ku_*}{fL} \tag{C.15}$$

has usually been used. The constants of integration A and B must then be considered to be functions of σ. It is now generally believed that under lapse conditions, the mixed layer depth is a more relevant length scale for the outer layer than u_*/f.

The Kazanski–Monin Equations make it possible to determine the surface layer wind profile and its direction when the geostrophic wind and z_0 are given.

Appendix D. Description
of the Planetary Boundary Layer Simulation Model

A computer diskette is provided with this publication. It is designed to be used with any PC type computer equipped with Windows 3.1 or Windows 95 operating systems. Before it can be run, it must be installed as described in the file README.TXT which can be found on the diskette. After installation, the program group can be run by clicking the icon labelled TURB & DIFFUSION. The group menu marked WINDOW will reveal two programs, the first of which is the planetary boundary simulation described in this appendix.

When the program is run, the screen displays a dynamically changing plot of the temperature, the dew point and the wind speed, as functions of height up to about two kilometers above the ground. The display marches in two-minute time steps, and the current local time is always visible. The initial distribution of the elements and the starting time are entered from one of the data sets which must be selected before starting the run. These data sets were derived from field programs and in some cases from routine upper-air observations.

Other features of the display include a dynamic plot of the terms of the surface heat balance equation as functions of time. There are also icons to show the relative altitude of the sun, the presence of free convection, the top of the free-convective layer, the presence of fog on the ground, and the location and types of clouds if and when they form.

Several options are available to the user. These include keeping a log of the temperatures of the air, ground, and foliage as well as the depth of free convection. The speed at which the model runs is dependent on the equipment. If it is too fast, the user can opt to renew the display at each time step, and to pause the evolution as often as desired.

The evolution of the boundary layer over a daily cycle is governed by a wealth of environmental variables and parameters, such as soil moisture and type, the free-atmospheric wind distribution, and air transparency, just to name a few. One of the important reasons for scientific study of turbulence is to develop a capability to assess the importance of environmental factors and their interactions in a decidedly nonlinear atmospheric system. Since controlled experiments on the real system are not possible, we depend on a realistic model. Experimentation requires the ability to "twist knobs" and observe the results; to poke the system in different ways and see how it reacts. Therefore, provision is made in this presentation of a list of 20 different important parameters that can be changed as desired before beginning a run. A description of each of these can be accessed by pressing a help button. Each has a default value from the data set.

A stand-alone version of the model (runable on DOS alone) is in the public domain and is listed in files that can be displayed using the File menu. The actual files from which the display is read can be concatenated and run on a QuickBasic interpreter or compiler.

This version of the program is a one-dimensional time-dependent model simulation of the planetary boundary layer. The model was developed at the Pennsylvania State University with support from the U.S. Environmental Protection Agency. Three-dimensional versions of the model are now being used in the Penn State – NCAR mesoscale model and the U.S. National Acid Deposition Model.

D.1 Architecture of the Model

The model implemented in the accompanying diskette is schematically pictured in Fig. D.1. It has an atmosphere consisting of 30 layers, denoted by the subscript i, each 100 m thick extending from a height of 10 m to 3010 m above the ground. (Only the lowest 2000 meters are displayed.) Within each layer, budgets are kept of each of the variables indicated: the internal energy, associated with the potential temperature θ (relative to the surface rather than sea level); the specific humidity q; the liquid water content q_l; and the components of the horizontal momentum u and v. These quantities are attributed to the midpoint of each layer: 60 m, 160 m, ... 2960 m. At the interfaces between each of these layers, the turbulent fluxes and the parameters on which they depend, shown in the figure, are calculated at every time step. A K-type parameterization is used except in layers that are undergoing free convection. The value of K is calculated from the ambient wind shear and Richardson number, as indicated in the figure.

Below these layers lies a surface layer 10 m thick within which Monin–Obukhov similarity is assumed, in place of a detailed resolution of the profiles. Within the surface layer a budget of the above variables is kept. Also within this layer lies a vegetative layer which interacts with the radiative fluxes, and through diffusion, with the air in the surface layer.

Beneath the surface layer is the ground surface, which receives the modified radiative fluxes beneath the vegetation and radiates upward as a black body, also reflecting the transmitted solar radiation in accordance with its own albedo. The ground surface also interacts diffusively with the surface layer air. Thus the ground and the vegetation behave analogously to a two-circuit electrical resistance network.

The complete specification of the boundary condition is carried out by the use of a two-layer ground-slab model of the soil devised by the author (1975), and which has come to be called the *force-restore* model. The lower slab of this model maintains a constant temperature (infinite heat capacity) and water content. The amplitude of the temperature variation in the upper surface and the times of

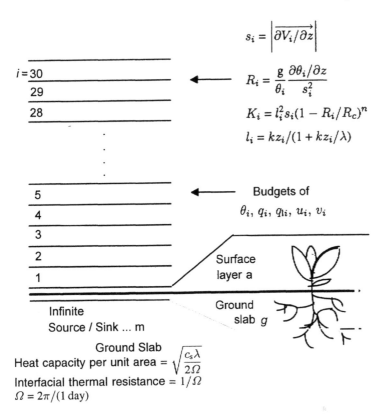

$$s_i = \left| \overrightarrow{\partial V_i / \partial z} \right|$$

$$R_i = \frac{g}{\theta_i} \frac{\partial \theta_i / \partial z}{s_i^2}$$

$$K_i = l_i^2 s_i (1 - R_i / R_c)^n$$

$$l_i = k z_i / (1 + k z_i / \lambda)$$

Budgets of

$$\theta_i, q_i, q_{li}, u_i, v_i$$

Surface layer a

Ground slab g

Infinite Source / Sink ... m

Ground Slab

Heat capacity per unit area $= \sqrt{\dfrac{c_s \lambda}{2\Omega}}$

Interfacial thermal resistance $= 1/\Omega$

$\Omega = 2\pi / (1 \text{ day})$

Fig. D.1 Architecture of the planetary boundary layer model discussed in this appendix. Budgeted quantities are represented by values at the center of the layer denoted by i. Variables associated with the fluxes are denoted by values at the top of the layer referred to by i

maximum and minimum are correctly determined when the heat capacity c_g per unit area of the upper slab is equal to $\sqrt{c_s \lambda / 2\Omega}$, where c_s is the soil heat capacity per unit volume, λ is the Fourier heat conduction coefficient, and Ω is the angular velocity of the Earth's rotation. The interfacial thermal conductivity (see Fig. D.1) must also be set equal to $1/\Omega$.

D.2 Surface Boundary Condition

The general features of the treatment of the surface boundary condition are illustrated in Fig. D.2. The surface heat balance equation (see (8.1)) is used to calculate the ground heat flux after introducing the solar radiation, the upward and downward streams of infrared radiation at the surface (see below), and the turbulent

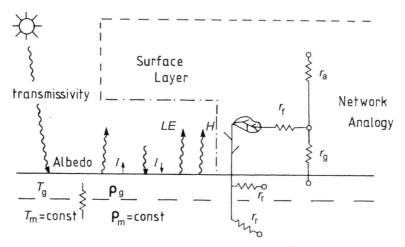

Fig. D.2 Schematic representation of the surface boundary used in the planetary boundary model

heat fluxes to the atmosphere (which are determined by resistance equations for the interfacial surface. Thus

$$G = (Q + q)(1 - A) - I_\uparrow + I_\downarrow - L_0 E - H_0 .$$ (D.1)

The ground surface temperature T_g is then calculated from the heat budget of the surface slab using c_g and the interfacial conductivity which were defined above;

$$\frac{dT_g}{dt} = \frac{G}{c_g} + \Omega \left(T_m - T_g \right) .$$ (D.2)

The moisture flux is handled by an electrical resistance network, as shown. All of the symbols are defined there except for ρ_x and ρ_w, which are, respectively, the maximum water content allowed by the type of soil, and the smallest soil water content that is accessible by the roots. The water transfer from the soil is strongly influenced by the presence of vegetation and processes controlled by it.

The treatment of the vegetation is an expanded version of the scheme used by Deardorff (1978). The main feature of this treatment is an improved model of the plant's mechanism for opening and closing of leaf stomata, and the use of the electrical resistance network analogy for handling the two paths of water movement from the soil to the surface layer. Closing of the leaf stomata increases the stomatal resistance r_f dramatically, and the determination of whether the stomata are open or closed is made on the basis of the solar radiation intensity and the plant's water budget. The latter is maintained by the model on the basis of the flow of water through the roots and the outflow through the leaves. If the soil water content is large, inflow is controlled by the root resistance r_r until the plant water content is saturated. In this case, the stomatal opening or closing is controlled by the solar radiation intensity. If the root flow cannot keep up with the loss through the leaves, the plant water content falls below a critical value, and the stomata close regardless of the radiation intensity.

The flow of water through the ground directly to the atmosphere is controlled by the slab water content ρ_g and the interfacial resistance r_g. The aerodynamic resistance r_a is the reciprocal of the product $C_q \overline{u}$ in (5.29), which, with this substitution, becomes an analogue of Ohm's Law.

D.3 The Free Convection Closure Scheme

When the ratio of the two velocity scales w_*/u_* begins to exceed the value 1/3, the model switches from K-type closure to a free-convection simulation. The underlying strategy of this type of closure is the familiar parcel method that is used by synoptic meteorologists to determine the energy available for release by convection and the height to which convection is able to penetrate. The first implementation of this type of closure was reported by Estoque (1968).

The implementation is best understood by referring to Fig. 7.5. The virtual potential temperature of the rising thermal is assumed to be that of the surface layer. The area indicated by the label + is proportional to the potential energy converted to kinetic energy by the rising parcel. The negative area, beginning at the level when the virtual potential temperature of the rising thermal is equal to that of the environment at that level, indicates the energy consumed by the parcel in rising through the inversion into the air above. The penetration is assumed to cease when the negative area is approximately 20% of the positive area.

At each time step, a small fraction of the air at each level is exchanged between the rising thermal column and its surrounding air layer. The fraction exchanged may vary with height as desired; usually a somewhat greater fraction is designated for the layers closest to the ground. As the air is exchanged, the enthalpy and other conservative properties are also exchanged in the same proportion. The fraction at each level is determined in such a way that the total net heat transferred from the rising column to its environment is equal to the heat that is transported out of the surface layer during the time interval. Thus, the energy is conserved and the system of equations is closed. Once the fraction is known at each level, the exchange of all of the other properties can be determined, and the requirements of conservation then permit the flux of each property into or out of the surface layer to be calculated.

A potential difficulty with this technique arises if the time steps are not kept very short. The amount of any property exchanged between the surface layer and the layers above depends on the surface layer value of that property, and the change that results from the exchange controls the amount of exchange in the following time step. To prevent oscillations that grow with time, a procedure has been developed for integrating the exchange equations analytically at all the levels over time steps of extended duration. The model runs with time steps of two-minute duration without any problems of computational stability.

D.4 Treatment of Cloud Formation

At each level, at each time step, a test is made to see if the layer is supersaturated. If this is found to be the case, the excess water vapor is condensed adiabatically and added to the liquid water content of the layer. Conversely, if the layer is unsaturated, liquid water is evaporated adiabatically to the extent that it can be provided from the layer's liquid water budget. Whenever a layer contains liquid water, a ###### icon is displayed at the appropriate height. In the case of the surface layer having liquid water, the symbol ====== is displayed instead. Layers with liquid water are always treated as black bodies when the infrared radiative transfer option (i.e. the default option) is activated.

An attempt is made to modify the transmissivity of incoming solar radiation for the effects of the stratus clouds described above, which are always assumed to cover the whole sky.

The rising thermals may also become supersaturated. When this occurs, the layers involved are displayed by the symbol *CUCU* at those layers that are affected. No liquid water budget is kept because the fraction of the area covered by cumulus clouds is not determined. For the same reason, the radiative transfer streams ignore the presence of convective clouds.

D.5 Treatment of Infrared Radiation

The infrared emissivity of a layer depends mainly on the precipitable water vapor content of the layer. A relationship derived from a precipitation chart published by Raethjen (1950) has been used for the PBL model. This relationship is shown in Fig. D.4. Since the water vapor content of each layer is always known, it is possible to estimate the emissivity at each level in the direction of an emerging stream.

In the absence of clouds, the infrared stream moving downward at any level is usually calculated from the temperature distribution as a function of water vapor path length looking upward. In practice, the intensity can be quite well calculated from the absolute temperature of the nearest layer and the emissivity of the path length to infinity. This is possible because the greatest part of the radiative flux is emitted by the adjacent layer, and most of the flux from the higher layers is absorbed in passing through the adjacent layer. The upward stream consists of two parts: the black body radiation emitted by the ground minus the portion absorbed in passing through the intervening layers, and the radiation emitted by the intervening layers. The absorbed fraction of the ground radiation is just the emissivity of the intervening layers, while the radiation emitted by the layer is determined in the same way as the downward stream, from the absolute temperature of the adjacent layer and the emissivity of the layer between the level and the ground.

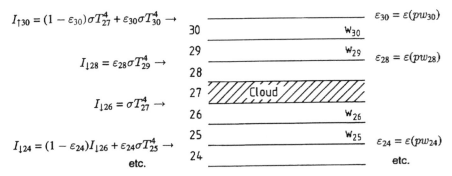

Fig. D.3 Examples of the infrared radiation stream calculation used in the planetary boundary layer model. Emissivities are calculated from the relation shown in Fig. D.4. The symbol pw_i stands for the total precipitable water between the subscripted interface and the appropriate black-body surface or the top of the atmosphere

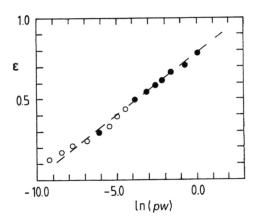

Fig. D.4 Relation between emissivity and precipitable water used in the planetary boundary layer model

If a layer of clouds lies below the level, the two streams are calculated in the same way as discussed above, except that the top of the cloud layer replaces the ground. If the level at which the two streams are to be calculated lies totally within the cloud, the upward stream is just the black body radiation emitted upwards at the temperature of the adjacent layer below the level, and the downwards stream is that emitted downwards by the adjacent layer above the level. If the level is between the cloud and the ground, or between two cloud layers, the upward and downward streams are calculated in the same way except that for the downward stream, the ground is replaced by the lower surface of the cloud.

As examples of this procedure, the upward stream emerging from layer 30 in Fig. D.3 is given by $\varepsilon_{\uparrow 30}$ in the illustration. In the equation, ε_{30} is the emissivity of the block comprising layers 28, 29, and 30. The downward flux at the top of layer 28 depends on the absolute temperature of layer 29 and the emissivity of the entire atmosphere above level 28. Since the moisture is undefined above level 30, it is assumed that the precipitable water above layer 30 is 15% of the initial water vapor content of the 30 given layers and that this quantity does not change with time.

The cooling rate calculated from this simple two-stream model is usually between one and two degrees centigrades per day, and agrees quite well with the prediction of more detailed models. When clouds are not present, the effects of radiation on the temperature change away from the ground itself are usually negligible. However, when clouds are present, radiative effects become quite important, as can be seen by experimenting with the model. When turbulence is present, fogs usually do not form. Instead, dew is deposited on the surface. Once a fog does form, perhaps as a result of the presence of a warm lake or brook, radiation becomes a dominant term and may cause the fog to thicken and spread rapidly to adjacent areas.

Appendix E. A Monte Carlo Smoke Plume Simulation

On the diskette provided with this publication is a Monte Carlo smoke plume simulation. Like the boundary layer model simulation, it can be run only from the program group that is generated by installing the compressed files on the diskette. If these have not previously been installed, refer to the file README.TXT on the diskette or the instructions in Appendix D. Once the installation has been made, one can run the program by clicking the TURB & DIFFUSION group and selecting SMOKE in the Windows Menu.

This program aims to provide an educational environment in which users can explore the effects of various environmental parameters on the shape and behavior of smoke plumes or individual cloud puffs emanating from a source that is either elevated or close to the ground. By using the *Settings* menu, the user can tailor this environment to suit his particular desires. Although the best available information about atmospheric processes has been employed throughout, the model has not been validated for operational use, and it should not be assumed that the results obtained from using it would satisfy present legislative requirements for licensing new enterprises.

Particles are issued sequentially from the source, the height of which may be designated by the user. The position of each particle is shown on each of two planes: (1) the projection on the xz plane and (2) the projection on the xy plane. (x is defined as the direction toward which the mean wind blows.) Each particle moves with the mean wind (which is a function of height) but has a velocity of its own relative to the mean wind. The latter is determined by a variation of the *random walk* process called the *drunkard's walk*. In the random walk process, the individual part of the velocity is redetermined at each time step by a random process that is uncorrelated with the previous time step. In the case of the drunkard's walk, the particle 'remembers' a portion of its previous velocity and has to substitute a randomly generated velocity to make up the remainder. The latter is determined such that the particle's velocity variances in the three directions satisfy the requirements of Monin–Obukhov and mixed layer scaling.

The rate of loss of memory is governed by the Lagrangian time scale, which is a function of the stability classification. The most unstable classifications have the longest time scales and tend to produce the largest and longest-lasting loops in the smoke plume. The model uses Hanna's formula (10.48) to determine the Lagrangian time scale.

The initial value of the individual velocity is determined in a similar way. Each initial value of the three velocity components of a newly emerging particle is determined by taking a fraction of the preceding particle's velocity component. To this is added a randomly generated supplement so as to ensure that it belongs to a statistical population appropriate to the selected environment and location. The amount of random supplement is determined by the rate of particle emission and by the Eulerian time scale, which is associated with the changes of velocity at a fixed location resulting from the passage of a 'frozen eddy' procession, as suggested by Taylor's hypothesis. This kind of time scale is related to the Lagrangian time scale, and is sensitive to the stability. The equation used in the model is (10.46).

Although Taylor's equation is not actually used as the basis for predicting the plume width, it demonstrates the importance of correctly estimating the velocity variances. For the surface layer, the vertical velocity variance is calculated from the Panofsky–Tennekes equation (8.18), which is based on Monin–Obukhov scaling. When this result begins to exceed that given by (8.20), we assume that the mixed layer scaling prevails. For the variances of the horizontal velocity components, (8.22) is used at all heights. Under stable conditions, h/L is treated as zero, but a small constant value is added to σ_u and σ_v so as to simulate meandering.

Built into the model is a simulation of the effects of stack emission buoyancy on plume rise following the empirical relationships suggested by Briggs. For the buoyant emission parameter Briggs used the symbol F, and this parameter can be set by the user. The default value of 5 is typical of the emission rates of many power plants. Users should have no difficulty in showing for themselves that the effect of this parameter is more important for low wind speeds than for higher ones.

Each of the active particles on the screen is represented by a number of memory locations in which are stored its velocity components, position, and other attributes, such as color, which can be used to identify individual particles and their displacements. Only a limited number of particles may be active simultaneously. To provide for an unlimited duration of the run, particles are recycled after they leave the screen.

Options displayed on the initial panel can be easily selected or changed from default values by moving the cursor and typing a new value. It is necessary to terminate the selected value with the *Enter* key to make it permanent.

One of the options provided on the initial panel is to observe the trajectory of individual particles. We have pointed out the vast difference between the appearance of a loopy plume and the motion of individual particles. This option can be run by changing the particle emission rate to 1. With the default conditions one gets a very loopy plume. However, when one looks at the trajectories of individual particles, they appear to be nearly straight lines. The loops result from the fact that successive particles move in quite different directions. Thus the shape of the plume is a snapshot of the history of the movement of a sequence of particles issued over a period of time, and nothing like the path of individual particles.

Some of the early models (particularly the ones associated with use of the Fick equation) fail to recognize the difference between plume dispersion and puff (or relative) dispersion. The latter is produced mainly by turbulent scales smaller than the diameter of the puff, while the former is produced mainly by the longest eddies in the turbulence spectrum. The large eddies affect the direction of successive puffs leaving the stack, while the small eddies determine the rate at which individual puffs spread after they are sent on their way by the large eddies. The importance of this difference can be appreciated by selecting a particle number between 2 and 49 – preferably the latter.

It will be noted from the initially displayed panel that the input parameters consist of the surface heating rate, the wind speed at five meters height, the surface roughness, and the height of the mixed layer. The stability class is not an input parameter, but instead is calculated from the vertical angular dispersion of the velocities as determined from the physical environment. The derived Pasquill–Gifford stability class is displayed at the top of the screen.

Each run involves the calculation of a very large sequence of pseudo-random numbers. The default procedure initiates this sequence with a pseudo-randomly generated seed using the output of the clock timer. Thus no two runs initiated in this way are identical. One of the options available to the user is to designate the initiating seed. By using this option the user can make identical runs repeatedly. This option is useful for demonstration purposes, where it may be useful to produce a particular feature predictably.

In the default graphical display, the length scale is the same in all directions. Tick marks are provided along the horizontal axes at 100 m intervals. The scale can be changed as desired by options on the initial panel.

When a particle impacts the ground, the event is recorded by incrementing a digit displayed under the point of impact. After the count reaches nine, additional impacts increment letters of the alphabet so that up to 36 impacts can be counted and displayed in each of the 80 locations. In this way it is possible to view the evolution of the integrated crosswind relative concentration as a function of downwind distance from the source.

An option is provided to designate the probability of deposition of a particle when it touches the surface. A choice of zero results in total reflection while a choice of 1 gives total absorption. When a particle impacts the ground a random number between zero and one is generated. If the number is less than the deposition probability, the particle is recycled (after counting it). Otherwise, its position is reflected from the position below the surface to a position above, and its previous vertical velocity is replaced with a randomly generated Gaussian-distributed vertical velocity consistent with the prevailing value of u_*. Thus after an impact with the surface, a particle loses all memory of its past.

Under default parameters, runs terminate automatically after 60 minutes of model time. This duration may be reset for shorter or longer times as desired. Runs may be aborted at any time by clicking RUN/STOP. When a run is terminated automatically, the smoothed crosswind integrated concentration is depicted graphically as a function of distance downwind. On the upper portion of the display, contours of the mean surface concentration on the xy-plane over the duration of the run are shown.

References

Arya, S. P., 1987: *Introduction to Micrometeorology*. Academic Press, 307 pp.

Blackadar, A. K., 1950: The transformation of energy by the large-scale eddy stress in the atmosphere. Meteorological Papers, 1(4), New York University, 33 pp.

Blackadar, A. K., 1957: Boundary layer wind maxima and their significance for the growth of nocturnal inversions. Bull. Amer. Meteorol. Soc., **38**, 283–290.

Blackadar, A. K., 1974: Implications of a simple two-layer model of the diabatic planetary boundary layer. Izvestiya Akad. Nauk, USSR. Atmos. & Oceanic Phys., **10**, 663–664 (English ed., Amer. Geophys. Union, **10**, 409–410).

Blackadar, A. K., and H. Tennekes, 1968: Asymptotic similarity in the planetary boundary layer. J. Atmos. Sci., **25**, 1015–1020.

Boussinesq, J., 1897: *Théorie de l'écoulement tourbillonant et tumultueus des liquides*. Gauthier-Villars, Paris.

Briggs, G. A., 1969: *Plume Rise*. U.S. Atomic Energy Commission Div. Tech. Inf., 81 pp.

Briggs, G. A., 1975: Plume rise predictions, in *Lectures on Air Pollution and Environmental Impact Analyses*, ed. Duane A. Haugen. American Meteorological Society, Boston, MA, pp 59–111.

Brunt, D., 1952: *Physical and Dynamical Meteorology*, 2nd Ed. Cambridge Univ. Press, 428+xxiv pp.

Businger, J. A., J. C. Wyngaard, T. Izumi, and E. F. Bradley, 1971: Flux-profile relationships in the atmospheric surface layer. J. Atmos. Sci., **28**, 181–189.

Caughy S. J., and S. G. Palmer, 1979: Some aspects of turbulence structure through the depth of the convective boundary layer. Quart. J. Roy. Meteor. Soc., **105**, 811–827.

Clarke, R.H., and G. D. Hess, 1974: Geostrophic departure and the functions A and B of Rossby number similarity theory. Bound. Layer Meteor., 7(3), 267–287.

Deardorff, J. W., 1978: Efficient prediction of ground surface temperature and moisture with inclusion of a layer of vegetation. J. Geophys. Res., **83**, 1889–1904.

Defant, A., 1921: Die Zirkulation der Atmosphäre in der gemässigten Breiten der Erde. Geografiska Annaler, **3**, 209–265.

Defant, A., 1926: Austauschgrösse der atmosphärischen und Ozeanischen Zirkulation. Annalen der Hydrographie und maritimen Meteorologie, **54**, Köppen Biheft, 12–17.

Dyer, A. J., 1967: The turbulent transport of heat and water vapour in an unstable atmosphere. Quart. J. Roy. Meteor. Soc., **93**, 501–508.

Dyer, A. J., 1974: A review of flux-profile relationships. Boundary-layer Meteor. **7**, 363–372.

Ekman, V. W., 1905: On the influence of the earth's rotation on ocean currents. Arkiv. Mathematik Astron. Fysik, Stockholm, 2(11), 1–52.

Ellison, T. H., 1955: The Ekman spiral. Quart. J. Roy. Meteor. Soc., **81**, 637–638.

Ellison, T. H., 1956: Atmospheric turbulence, in *Surveys in Mechanics*, ed. by G. K. Batchelor and R. M. Davies, Cambridge, pp 400–430.

Ellison, T. H. 1957: Turbulent transport of heat and momentum from an infinite rough plane. J. Fluid Mech., **2**, 456–466.

Estoque, M. A., 1968: Vertical mixing due to penetrative convection. J. Atmos. Sci., **25**, 1046–1051.

Garratt, J. R., 1992: *The Atmospheric Boundary Layer*. Cambridge Univ. Press. 316+xvii pp.

Geiss, H., K. Nester, P. Thomas, and K. J. Vogt, 1981: *In der Bundesrepublik Deutschland experimentell ermittelte Ausbreitungsparameter für 100 m Emissionshöhe*. Ber. Kernforschungsanlage Jülich, Nr. 1707, 29 pp. (Also Ber. Kernforschungszentrum Karlsruhe, Nr. 3095.)

Gifford, F., 1968: Diffusion in the lower layers of the atmosphere, in *Meteorology and Atomic Energy*, ed. D.H. Slade. U.S. Atomic Energy Commission, Washington, D.C., pp 65–116.

Hanna, S., 1981: Lagrangian and Eulerian time-scale relations in the daytime boundary layer. J. Appl. Meteor., **20**, 242–249.

Haugen, D. A. (ed.), 1973: *Workshop in Micrometeorology*. Amer. Meteor. Soc., Boston, 329 pp.

Heisenberg, W., 1948: On the theory of statistical and isotropic turbulence. Proc. Roy. Soc. (A), **195**, 402–406.

Hinze, J. O., 1959: *Turbulence*. McGraw Hill, New York, 586 pp.

Hojstrup, J., 1982: Velocity spectra in the unstable boundary layer. J Atmos. Sci., **39**, 341–356.

Kaimal, J. C., J. C. Wyngaard, Y. Izumi, and O. R. Cote, 1972: Spectral characteristics of surface-layer turbulence. Quart. J. Roy. Meteorol. Soc., **98**, 563–589.

Kaimal, J. C., J. C. Wyngaard, D. A. Haugen, O. R. Cote, Y. Izumi, S. J. Caughey, and C. J. Readings, 1976: Turbulence structure in the convective boundary layer. J. Atmos. Sci., **33**, 2152–2169.

Kampé de Férier, M. J., 1939: Les fonctions aléatoires stationaires et la théorie statistique de la turbulence homogène. Ann. Soc. Sci. Brux., **59**, 145.

Kazanski, A. B., and A. S. Monin, 1961: On the dynamical interaction between the atmosphere and the earth's surface. Bull. Acad. Sci., USSR, Ser. Geophys., Nr. 5, 786–788.

Kolmogorov. A. N., 1941: The local structure of turbulence in an incompressible fluid for very large Reynolds numbers. Dokl. Akad. Nauk SSSR, **30**, 301–305 and **32**, 16–18.

Kraus, E. B., 1972: *Atmosphere – Ocean Interaction*. Clarendon Press, Oxford., 275 pp.

Kraus, Helmut, 1970: Die Energieumsätze in der bodennahen Luftschicht. Ber. Deutsch. Wetterdienst (Offenbach), Nr. 117, 43 pp.

Lettau, H. H., and B. Davidson, 1956: *Exploring the Atmosphere's First Mile*: Vol. 1, *Instrumentation and Data Evaluation*; Vol. 2, *Site Description and Data Tabulation*. Pergamon Press, New York., 578 pp.

Lettau, H. H., 1950: A re-examination of the "Leipzig wind profile" considering some relations between wind and turbulence in the friction layer. Tellus, **2**, 125–129.

Lettau, H. H., 1957: Windprofil, Innere Reibung und Energieumsatz in den unteren 500 m über dem Meer. Beitr. z. Physik der Atmos., **30**, 78–96.

Lumley, J. L., and H. A. Panofsky, 1964: *The Structure of Atmospheric Turbulence*. Interscience Publishers, New York, 239+xi pp.

Monin, A. S., and A. M. Obukhov, 1954: Basic laws of turbulent mixing in the ground layer of the atmosphere. Trans. Geophys. Inst. Akad. Nauk, USSR, **151**, 163–187.

Monin, A. S., and A. M. Yaglom, 1971: *Statistical Fluid Mechanics. Vol. 1*. English Ed. MIT Press, Cambridge, MA, 769 pp.

Monin, A. S., and A. M. Yaglom, 1975: *Statistical Fluid Mechanics. Vol. 2*. English Ed. MIT Press, Cambridge, MA, 300 pp.

Navier, M., 1827: Mémoire sur les Equations générales de l'Equilibre et du Mouvements des Fluides. Mem. de l'Acad. d. Sci., **6**, 389.

Nieuwstadt, F. T. M., 1984: The turbulent structure of the stable, nocturnal boundary layer. J. Atmos. Sci., **41**, 2202–2216.

Nieuwstadt, F. T. M., and H. van Dop, 1982: *Atmospheric Turbulence and Air Pollution Modelling*. Reidel, Dordrecht, 358 pp.

Onsager, L., 1945: The distribution of energy by turbulence Phys. Rev. **68**, 286.

Panofsky, H. A., 1963: Determination of stress from wind and temperature measurements. Quart. J. Roy. Meteor. Soc., **89**, 85–94.

Panofsky, H. A., A. K. Blackadar, and G. E. McVehil, 1960: The diabatic wind profile. Quart. J. Roy. Meteor. Soc., **86**, 495–503.

Panofsky, H. A., H. Tennekes, D. H. Lenschow, and J. C. Wyngaard, 1977: The characteristics of turbulent velocity components in the surface layer under convective conditions. Bound. Layer Meteor., **11**, 355–361.

Panofsky, H. A., and J. A. Dutton, 1984: *Atmospheric Turbulence; Models and Methods for Engineering Applications*. Wiley and Sons, 397+xix pp.

Pasquill, F., and F. B. Smith, 1983: *Atmospheric Diffusion*, 3rd Ed. Ellis Horwood, Ltd., 437 pp.

Paulson, C. A., 1970: The mathematical representation of wind speed and temperature profiles in the unstable atmospheric surface layer. J. Appl. Meteor., **9**, 857–861.

Phillips, P., and H. A. Panofsky, 1982: A re-examination of lateral dispersion from continuous sources. Atmos. Envir., **16**, 1851–1859.

Prandtl, L., 1925: Über die ausgebildete Turbulenz. Zeitschr. für angew. Math. u. Mech., **5**, 136.

Prandtl, L., 1932: Meteorologische Anwendung der Strömungslehre. Beitr. Phys. Atm., **19**, 188–202.

Priestley, C. H. B., 1959: *Turbulent Transfer in the Lower Atmosphere*. Univ. Chicago Press, Chicago, 130 pp.

Raethjen, P., 1950: *Kurzer Abriss der Meteorologie dynamisch gesehen. Teil II, Wärmehaushalt der Atmosphäre*. Geophysicalische Einzelschriften, Geophysikalisches Institut der Universität Hamburg, 103–152.

Reid, J. D. 1976: Markhov chain simulations of vertical dispersion in the neutral surface layer for surface and elevated releases. Bound. Layer Meteor., **16**, 3–22.

Richardson, L. F., 1920: The supply of energy from and to atmospheric eddies. Proc. Roy. Soc., A, **97**, 354–373.

Rossby, C. G., 1941: The scientific basis of modern meteorology, in *Climate and Man, 1941 Yearbook of Agriculture*. Department of Agriculture, Washington, pp. 599–655.

Rossby, C. G., and R. B. Montgomery, 1935: The layer of frictional influence in wind and ocean currents. Papers in Phys. Oceanog. and Meteor. Mass. Inst. Tech. and Woods Hole Oceanog. Inst., 3(3), 101 pp.

Rotta, J. C., 1972: *Turbulente Strömungen*. B. G. Teubner, Stuttgart, 267 pp.

Schmidt, W., 1925: *Der Massenaustausch in freier Luft und verwandte Erscheinungen*. Henri Grand, Hamburg.

Sellers, W. D., 1962: A simplified derivation of the diabatic wind profile. J. Atmos. Sci., **19**, 180–181.

Shao, Y., and J. M. Hacker, 1990: Local similarity relationships in a horizontally inhomogeneous boundary layer. Bound. Layer Meteor., **52**, 17–40.

Schlichting, H., 1965: *Grenzschicht-Theorie*. Braun, Karlsruhe, 736 pp.

Starr, V. P., 1948: On the production of kinetic energy in the atmosphere. J. Meteor., **5**, 193–196.

Starr, V. P., 1951: Application of energy principles to the general circulation, in *Compendium of Meteorology*. ed. by T. F. Malone, Amer. Meteor. Soc., Boston, pp. 568–574.

Stokes, G. G., 1845: On the Theories of the Internal Friction of Fluids in Motion. Trans. Cambridge Philos. Soc., **8**, 1240.

Stull, R. B., 1988: *An Introduction to Boundary Layer Meteorology*. Kluwer, Dordrecht, 666 pp.

Swinbank, W. C., 1963: Long-wave radiation from clear skies. Quart. J. Roy. Meteor. Soc., **89**, 339–348; discussion, **90**, 488–493.

Taylor, G. I., 1915: Eddy motion in the atmosphere. Phil. Trans. Roy. Soc. (London), Ser. A, **215**, 1–26.

Taylor, G. I., 1922: Diffusion by continuous movements. Proc. London Math. Soc., **20**, 196–212.

Taylor, G. I., 1931: Effect of variation in density on the stability of superposed streams of fluids. Proc. Roy. Soc. (London), Ser. A, **132**, 499–523.

Taylor, G. I., 1932: The transport of vorticity and heat through fluids in turbulent motion. Proc. Roy. Soc. London, Ser. A, **135**, 685–702.

Taylor, G. I., 1935: Statistical theory of turbulence, Part I. Proc. Roy. Soc. London, Ser. A., **151**, 421–444.

Tennekes, H., and J. Lumley, 1972: *A First Course in Turbulence*. MIT Press, Cambridge, MA, 300 pp.

Thomas P., and K. Nester, 1985: Experimental determination of the atmospheric dispersion parameters at the Karlsruhe Nuclear Research Center for emission heights of 60 and 100 m. Nuclear Technology, **68**, 293–310.

Willis G. E., and J. W. Deardorff, 1974: A laboratory model of the unstable planetary boundary layer. J. Atmos. Sci., **31**, 1297–1307.

Weizsaecker, C. F. von, 1948: Das Spektrum der Turbulenz bei grossen Reynoldsschen Zahlen. Z. Physik., **124**, 614–627.

Woods, J, D., and V. Strass, 1986: The response of the upper ocean to solar heating. II The wind driven current. Quart. J. Roy. Meteor. Soc., **112**, 29–42.

Wyngaard, J. C., and O. R. Cote, 1971: The budgets of turbulent kinetic energy and temperature variance in the atmospheric surface layer. J. Atmos. Sci., **28**, 190–201.

Wyngaard, J. C., and O. R. Cote, 1972: Cospectral similarity in the atmospheric surface layer. Quart. J. Roy. Meteor. Soc., **98**, 590–603.

Yamamoto, G., 1959: Theory of turbulent transfer in non-neutral conditions. J. Meteor. Soc. Japan, **37**, 60–70.

Index